THE DARK STUFF

To Derek
in the hope you'll keep
digging!

THE DARK STUFF

Stories from the Peatlands

Donald S. Murray

Donald S. Murray

Illustrations by Douglas Robertson

B L O O M S B U R Y
LONDON • OXFORD • NEW YORK • NEW DELHI • SYDNEY

To Roel and Marleen, with thanks for help
To Maggie, with love

BLOOMSBURY WILDLIFE
Bloomsbury Publishing Plc

50 Bedford Square, London, WC1B 3DP, UK

BLOOMSBURY, BLOOMSBURY WILDLIFE and the Diana logo are trademarks of Bloomsbury Publishing Plc

First published in Great Britain 2018

A catalogue record for this book is available from the British Library.

Library of Congress Cataloguing-in-Publication data has been applied for.

ISBN: HB: 978-1-4729-4275-3
ePub: 978-1-4729-4278-4
ePDF: 978-1-4729-4277-7

2 4 6 8 10 9 7 5 3 1

Illustrations by Douglas Robertson

Typeset in Bembo Std by Deanta Global Publishing Services, Chennai, India
Printed and bound in Great Britain by CPI Group (UK) Ltd, Croydon, CR0 4YY

Contents

For a moment he thought of returning lest he should get lost in the space of the moor, but some daring instinct, some sense of adventure, was urging him onwards from the houses anchored to their familiar earth. It was as if he was Columbus setting off into a new world with inadequate maps, charts that showed only dimly and tentatively unknown seas and unknown islands.

On the Island, Iain Crichton Smith

PASSING

(from an Irish proverb)

The child arrives, weighty as a creel filled with peat,
for many years, providing warmth and heat,
till he or she departs, light and insubstantial as white
wisps of smoke rising from a fire banked up late at night.

PART ONE

Rùsgadh – Turfing

CAMOUFLAGE

The eye can be deceived by flowers
bright upon the surface. They bloom
for – say – a season, conceal the toughness of the limbs
digging deep below. Each plume's
a small deception, as men find out when they exhume
and unearth heat from damp and cold
for heather knots as tight and hard as briar,
each flare of purple like a rose
spilling out from thorns
disguising the resilience
of how its grip clings to soil and stone
through the bushwhack of each storm.

CHAPTER ONE

Fraoch (Scottish Gaelic) – Heather

I grew up on a green spit of land, with acres of emptiness on either side.

Behind our house on the Isle of Lewis there was the width of the Atlantic. Only around three-quarters of a mile separated it from the back of our byre, a tractor trail running down to the shoreline through the centre of our narrow strip of croftland. Green stalks of oats, crops of potatoes and turnips guarded you along the way. A cow or some sheep peered as you passed. Looking out from a flat, stony edge of land, one we called the *cladach,* differentiating it from the green, fertile soil of the *machair* a few miles to the north, there was little to be seen – a plinth of rock called Sgeir Dhail (Dell Rock) on which a herd of seals sometimes settled, making known their presence by a cacophony of loud, assertive honks; the curl of the Butt of Lewis with its red sandstone lighthouse; a few gulls and gannets, shags and cormorants, the last slinking below seawater, all sleek and black. There were times I would stand on tiptoes and dream I saw Newfoundland in the distance far to our west, its place names like a litany of prayer I had stumbled across in a book. I would pretend I could pinpoint where they were with my finger, stabbing into the vacancy:

'Old Bonadventure … Kettle Cove … Heart's Content … Swift Current …'

Their names were all so much more poetic and exotic than the place names of the villages of Ness all around me. 'South Dell … Cross … Skigersta … Swainbost … Adabrock …' Only the township of Eoropie in the far north possessed the breadth and expansiveness that I heard in those places on the far side of the Atlantic, as if somehow it was less cramped and confined than its appearance might suggest, holding a continent within its borders, a vowel or two misplaced in its search for Europe's vast entirety.

There were times I needed that sense of belonging, of not being isolated from the world. Our house faced a similar kind of emptiness to the one I witnessed in the width of the sea. The only difference was that it retained a flat calmness, even on days when the wind stirred up the ocean. Only one or two breakers broke the surface. Closest to our home, Beinn Dail (Dell Mountain) was like *The Great Wave off Kanagawa*, created by the Japanese artist Hokusai in a wood print in the early nineteenth century. It, however, was more of a rogue wave than a tsunami and one that would do little damage to either man or the environment. Its tiny 170 yards in height were unlikely to threaten either the empty land below or South Dell in the distance if it ever fell or toppled, casting rock or peat from its summit. Another hill, Mùirneag, was the equivalent of Mount Fuji in that portrait, peering over Beinn Dail's edge. Again, at just 271 yards in height, it was miniature, only attracting attention because of the horizontal nature of much of its surroundings. Looking up from the depths of a bog, it seemed large and looming, an 'Everest' dominating the scene.

Yet despite its lack of dizzy, death-defying heights, our moor was an intimidating landscape. There were miles when little could be seen but chains of bogs and still black pools; low hills; outcrops of heather; obstacles to anyone who wished to progress across its acres. When I tried once or twice to walk across to the village of Tolsta on the opposite side of the island, that motionless, flat landscape which at first glance seemed short in terms of distance became huge and as expansive as the ocean. There were occasions, too, when I might roll and reel, as if I were on a sea voyage. Feet could slip and slide, tumbling towards a pool of water. The heather ridges seemed to mount up before me, crest upon crest, forming insurmountable barriers. There was a maze-like quality about walking across its breadth. You could easily stumble across your own steps, crisscrossing the moorland in endless, decreasing circles. This meant that people could become lost, either in mist or blizzard. The last was a fate that occurred, for instance, to three young men, Allan Campbell, John Macritchie and Angus Morrison, way back in December 1883. They were found some six miles from their home in the village of Lionel, dead from cold and exhaustion.

It is for this reason that, wherever you go on the Lewis moor, there are often cairns of stone or slabs of rock upended, markers for a route that it might be advisable for a person to take or even to indicate where a person might gain some perspective on his place and position in that austere world. My great-grandfather, nicknamed 'Stuffan' after Stephen in the Bible, hefted a pinnacle of stone on his back and placed it deep within the peat on the Dell moor.* There were others who did this too, guides and navigators who provided pinpoints of light, pinnacles of Lewisian gneiss by which people could see their way across this nondescript landscape. They were also – according to a friend of mine – likely to have laid paths too, especially in areas that were well used by them, between the stone, tumbledown walls of sheilings, small patches of pasture. Nowadays, the paths stretch unseen, concealed by moss or heather.

There were human dangers on the moor too. Even as relatively late as the last years of the First World War, when 'so many of our dear ones and best manhood [were] laying down their lives for King and Country', the crofters of my native village South Dell petitioned Sheriff Substitute Dunbar in the Burgh Court in Stornoway about the actions of a man

* There were so many of the same names in my native district that virtually all men – and some women – were given nicknames. Many of these were plucked from the pages of the Bible, with other people in our village known as 'Isaac', 'Timothy' and 'Thomas'.

called Norman Macdonald. This individual was apparently employed as a gamekeeper in Galson Farm – at that time the only large working farm that bordered the district. Claiming he was a 'source of danger', they noted he had shot a gun in the direction of a young lad called William Murray – someone I knew as an old man and one of the church elders – for 'having a rabbit in his possession which the dog he had with him killed accidentally'. Two weeks before that, he had 'pointed his gun point blank [at] another man from this township on the public road'. He had also threatened people and 'been convicted for assaulting a young boy on the shoreline'. It was little wonder that the 'very children [were] afraid for their lives', fearing that 'this individual will meet them when they are herding the cattle or looking after their sheep on the moor'. They followed a maze-like path to avoid Macdonald, looking out for his presence looming in their direction as he came across the moor.

Yet these dramas were long past by the time of my childhood. As I looked out the window of my home, there were times instead when this landscape seemed dull and mundane. As Seamus Heaney writes in his poem 'Bogland', there were no 'prairies' in the moorlands of districts like Ness; it was far too small and limited to be compared with the vast flatness of that American landscape. The eyes of islanders have little to gaze outwards upon as they stare at the moor. Instead, sight dwindles downwards to focus on the tiny flowers that cluster around the feet. There are stretches of bog cotton that look like ripples on waves as they flap in the wind; miles too of the different varieties of heather* – from bell heather to

* There are, of course, a large number of different kinds. The heath family comprises about 40 genera with more than 1,400 species. The largest of these is the *Erica* genus, which contains heather among its members. *Erica* consists of roughly 860 species, with most of them distributed throughout Southern Africa, though there are some in the Mediterranean and Atlantic areas. The

cross-leaved heath and ling heather – stretching outwards to the horizon. There are varieties of plants like brightly coloured starfish gleaming on the surface of the moor, as if washed up on its dampness after some imaginary tide has receded. Bogbean and bog cotton twinkle among bleakness, capturing my attention as I head out on my 'walk' – a word that seems inadequate when one considers the bouncing quality, like a trampoline, the land often had below the feet. And then there was the way flowers like bog asphodel flamed yellow or orange, glistening in the poor soil. There were days when the moorland seemed a patchwork of colours, each one threading into another, a mingling of purple, chocolate brown, even rosy shades. Even water had an unusual clarity, largely because it was 'oligotrophic' or low in nutrients. This led to very little moss or other kinds of plant life forming on its surface.

These were times, too, when sunlight glinted on a loch, bright perhaps with water lilies, white and delicate at the edge of peat-shaded depths, or when wild iris (*sealastair*) decorated a stream with its yellow or orange flags. Even these waters were a source of mystery to me. For a moment, you could see them bright and clear as you walked along the edge. Then they would disappear below the surface, gurgling and splashing as they vanished beneath a layer of earth and short, cropped grass, an unexpected green rim on a khaki-coloured landscape. As a boy, I loved to walk along this, dawdling above the rumble of water below my feet, conscious that the stream was still there, my hearing and sense of touch able to navigate the direction in which it was racing, discovering I was right a few moments later when its cover was 'blown' and I could see it once again, as fresh and clear as ever.

---

majority of plants in the Barvas moor – like most of the Highlands of Scotland – would have been *oxylophytes*, adapted to growing in sphagnum moss. They include, among other plants, dwarf birch, low willows, sundew and cotton grass.

One of the visiting ministers to our parish, Rev. MacSween, a former fisherman from the Isle of Scalpay in Harris, conjured up the memory of that walk one time in his sermon in our small village church. He described it in much the same way as I 'saw' it – how a stream could seem loud and clamorous one moment and then fade away the next, vanishing under turf. 'Only its echo can be heard,' he said, 'as if it is only the memory of a stream. And then we see it again, sharp and clear as ever.'

'Our faith can sometimes be like that,' he continued. 'Sometimes its call can be strong within us. Then its presence all but disappears. A short time afterwards and we can see it by our side again, as sharp and pure as ever. Our sense of its existence as strong as it was when we first encountered it.'

In locations where the stream flows clear and open, the difference between 'land' and 'water' is distinct and well-defined. The edge of each gush and torrent is definite and clear-cut, as if it has been drawn into the earth by the sharp edge of a spade. This is not the case throughout the moor. There are many areas where there is no such contrast, where the two blur into one. These are the moorland 'bogs', a word derived from the Gaelic word for 'soft', *bog*, one that describes these parts of the world correctly and accurately, the way the soil slithers in your fingers as you seek to clutch its substance in your hand. Step carelessly on its surface and it is easy to break through from one element into another, solidity turning into liquid, firmness becoming fluid.

I recall how it transformed below my feet on more than one occasion, most especially the evening I lagged behind some of my boyhood friends on the edge of the village that faced out to the moor. In the half-light of day, these village boys were all blind to where I was going. No one was standing either at household windows to observe how I had stumbled into a green patch of land at the edge of a fence, astonished at how the damp ground swelled up to open and invite me, ushering my body within. No one heard me either; certainly not my frantic cries as the bog sucked in my wellingtons, swallowed up my legs. My contemporaries, after all, were lost in a discussion about Jimmy Johnstone, a Celtic footballer who possessed the knack of evading tackles, keeping the ball at his feet despite the outstretched boots of his opponents and their forlorn attempts to overwhelm his balance by a quick nudge of their shoulders. Clearly this topic was far more important to my friends than the clumsy straggler trailing behind their footsteps, one who had been unable to stay upright.

It was at this point I probably needed some of the faith the Rev. MacSween had spoken about. In the dimming of the light, I felt that my light was also dimming. I was sinking deeper, bog creeping above my waistline, convinced there was no bottom to the earth I was no longer standing on, that even if there was, my feet would never find it, not until – at least – the water had crept over my nose and mouth. In a frenzy, I imagined a search party being formed by the inhabitants of the village – Doilidh Dhodu, Iain Dido, Domhnall Thormoid, my dad – and finding my dead body on the outskirts of the village, looping a rope around my chest and drawing me out of the water's grasp. The boys of the village would be weeping. 'We didn't hear him shouting,' they would try to explain. 'We were talking about Jimmy Johnstone at the time.'

Yet it all never happened. Somehow or other, my soles touched – what might have been – bottom. Gravity slowed and halted its pull. As the bog oozed its way to my chest, I discovered a way of saving myself. I dug fingernails into earth, gripping grass and soil as tightly as I could. Slowly, I drew myself into what remained of the daylight, inching my way upwards from the clutch of the quagmire. A moment later and I was pushing my arm against firm ground. Leaning forward, I heaved my chest onto the earth, dragging myself from the wetness. My waistline was clear now. Soon, I was on a spot where my feet could find firm land again. My wet jeans swished and swirled, already stiffening in the coldness of the fading light as I tried to regain what was left of my balance and dignity, rushing off to join the others in their discussion of the merits of certain Celtic players, telling them, too, of how the earth had given way below my feet. They chuckled at what had happened, sounding not unlike the geese which cackled – like witches on broomsticks – as they flew across the twilight overhead.

'Trust you!' they said.

There were other forms of life that were not so fortunate in their struggles with bog and water. Sheep, especially yearlings, frequently lost that fight. (It seemed regularly to occur to the lambs my dad gifted me each spring. The notion that they were *caora Dhòmhnaill* or 'Donald's sheep' apparently compelled them to commit a form of suicide on the open moor the following autumn, a fate the ones my brother had been given mysteriously avoided.) My dad would be often disturbed when working on the loom, someone telling him that one of the flock had been drowned near, say, Asmigarry or Allt Leanabhat. He would nod his head and, as soon as it was convenient, take out a pair of thick woollen socks and wellington boots, tug them on his feet, and head in the direction of the moor. He would clutch a spade as he strode towards the stream where the animal lay, aware that as soon as he reached that scabrous, peat-straggled corpse, he would have to bury the body. There were reasons why this was a priority in our area. Once or twice, residents of Ness had suffered from the illness hydatid, caused by a tapeworm passed to them through the hair of household dogs that had discovered the unburied carcasses of sheep out on the moor. Unless it was treated quickly, this contact could kill, making its way into the human brain, lung or heart. As a result, men like my dad went out to dig graves for their animals as swiftly as they were discovered, laying their carcasses down as deeply as it was possible to do.

A few were not as scrupulous. Pressed too much by time, too little by conscience and a sense of duty to others, one of the more idle crofters in the area might wrap up the dead sheep in a plastic bin bag. At other times, they would not trouble themselves to do even this, leaving them unburied on the moor. Fleece and flesh were quickly shorn by the bite of cold weather and maggots, leaving a skeleton like the way Norman MacCaig describes that of a hind in his poem 'So Many Summers', its immaculate geometry resembling a boat tied up on the moor. Their skulls would be gradually exposed too. There was one boy among my near-contemporaries in the village who used to collect, inside a fish box, the horns that curled out from the sides of their skulls, gathering them too when they became snapped off after being snarled in the wire of a village fence. Occasionally he would take them out, spinning them round his fingers or dangling them from his pockets, Colt 45s that allowed him to play the role of a cowboy swaggering down the village. Gunfight at South Dell Post Office. Bushwhacking at the Kirk.

These were decades in which – Wild West-like – large flocks of animals roamed the South Dell moor. Hardy Blackfaces for the most part, they spent a large segment of the year out there, obtaining a thin, bare sustenance on its meagre acres, roods and perches – to borrow the language my older neighbours were taught in primary school. During winter, they grazed on croftland, their absence from it in the summer months allowing the fields time to recover and grow. Within these fences, the grass was also transformed into hay. This was also a necessity for the animals. If the cold months of the year were exceptionally harsh, they could be ushered into barn or byre to be fed. Swathes of hay might also be provided outside too – if the grass was cropped too short to give food to the animals.

Sheep were also the reason I came to know the moor for the first time. I recall going out there with a neighbour and great-uncle, Alex Smith, when I was around eight or nine. He was in his seventies, kitted out in dungarees, old-style denim jacket and cloth cap as he paced across its terrain with a walking stick clenched in his hand. The stick served two purposes. Not only did it assist him in his journey, it could also be used to hook a sheep by the horns or neck, drawing the animal towards him. His specific task that day, however, was more awkward than that – to look after me on my first venture on the moor. He would point out to me stretches that looked as if they had been daubed with a gaudy shade of green, a sign that they were soft and treacherous. He would show me shortcuts, the easiest ways for people to scale a peat bank, get across a stream. In short, he was chosen to accompany me as he was someone with a vast knowledge of the moor, able to point out all its landmarks and hazards, provide me with a sense of its size and scale, the island at its broadest span.

We did this on the day that the parish's most important citizens were returning. My father and the other men of the village had gone out to fetch the sheep from the moor, bringing them back to a small fence – or fank – that was not far from the community's edge. There they were to be clipped free of wool, a fresh coat of paint daubed on their backs, and sometimes to be dipped in water designed to kill and destroy the ticks and other creatures that clustered on their skin. They would be dosed with medicine, too, a black-tipped 'gun' firing pills and capsules down their throats. As they were gathered, it was Alex's role to keep a quiet eye on the flock, making sure that none of them bolted and escaped. It was my task to run, flap my arms and yelp like a collie, joining in to chase any that had the will and temerity to break free.

Later I discovered this double act was not an unusual one in the Highlands and Islands at that time. In Neil M. Gunn's novel *Young Art and Old Hector*, there is a similar relationship to the one I enjoyed with Alex between an elderly man and a young boy. The older man is a model of moderation and self-control, the child quick and impulsive. Just as Alex passed on his knowledge to me, Hector passes his awareness of the Highland environment on to Art. One can only wonder whether my companion ever obtained anything in return from me. From the freshness of my view of the countryside around us, did he, like his fictional counterpart, obtain a return of: 'that early rapt wonder, which had been lost for many years, opened its own eyes within [Hector] as he once more beheld the world'?

There were many reasons for Alex's awareness of that landscape. He was one of that generation of people who had spent much of their lives out on the moor, at first the one that was connected to his original village of Eoropie, later the much larger one that 'belonged' to South Dell. With his family, he would probably have gone out every summer to one of the *àirighean* (sheilings) that dotted the Ness moor. Most were little more than shacks, a tumbledown of stone or turf barely watertight or wind-tight and only habitable during the warmer months of the year. Small, dark Hebridean cattle would graze there, sipping from the lochs and streams that ribbon the moor. Children would have been able to run around the building until darkness brought an end to the long hold of dusk upon the landscape. This was their holiday time, a period, too, when their mothers might escape the back-breaking monotony of croftwork, leaving their men behind to weed potatoes, thin out a field of turnips. It is probably – or so I have been told – one of the last remnants of the ancient transhumance tradition that used to be found throughout northern Europe and beyond at one time, people shifting their flocks to take advantage of new fresh grazing somewhere else.

Ghost stories were probably told within the sheiling's rough-hewn walls. Illicit meetings probably took place, far from the restrictions of village life, particularly among those in their late teenage years or early twenties. This was life on the margins of island communities, with a general state of mayhem being permitted that would not be allowed within the more rigorous walls of home. There is a story from the Isle of Benbecula – one of a number of such stories found in the Western Isles, called *àirigh na h-aon oidhche* or 'sheiling of the one night'– that illustrates this. It is about a married man with the name of MacPhee who went out with six male companions to build a sheiling on that island's moor. When they had finished building, they decided to sleep there that night, using heather to bed down upon. It was at this point that one of MacPhee's friends declared, 'I wish our girlfriends were with us here.'

MacPhee must have smiled at his words, content with the thought that – unlike the others – he had a wife at home. Moments later, however, seven women – the eighteenth-century equivalents of Monroe, Harlow and Bardot, perhaps – appeared out of the twilight. They whirled and pirouetted around the sheiling before slipping down beside the men, kissing them and running long, graceful fingers down their backs.

Only MacPhee refused. He shifted away from the others, heading in the direction of the dwindling fire. When his supernatural equivalent of Sophia Loren circled him, he shook his head and refused her offer, muttering something about his wife at home. Glancing back at the others, he noticed that the faces of his companions were changing colour, their features drained of blood.

'Come on!' he yelled. 'Get out of here!'

He was the only one who did so, racing out of the doorway towards home. The others were either too far gone or too enamoured with the company of their bed-partners to follow. As he rushed across the moorland, the ghostly woman followed him, calling out his name, beseeching him to turn back. Her chase only ended when he set his dog upon her. She turned back again towards the sheiling, disappearing once more into the dusk's immensity.

The story doesn't end there. When MacPhee reached the first house on the edge of the moor, he found seven plates filled with milk lying on the ground outside. Suspicious of their presence, he did not reach down towards them, but watched instead as his dog lapped up their contents. A few moments later and the animal was dead. The same proved true of his friends when he journeyed out with others to the sheiling the following morning – each one a corpse lying in a shroud of heather.

And so, through both exploring its geography and tales like these, as much warnings as ghost stories, the likes of Alex learned to 'know' both its emptiness and scale, recognising its contours almost as well as the face he used to lather with a shaving brush, sharpening the cut-throat razor employed upon it on a dark leather belt. He might have been one of those who relished his time on the moor, looking back on it in the way Gaelic-speaking exiles were said – falsely – to have viewed life there, the anonymous 'Canadian Boat Song' of the mid-nineteenth century summing up their *cianalas* (longing to return) in the words:

From the lone shieling of the misty island
Mountains divide us, and the waste of seas –
Yet still the blood is strong, the heart is Highland,
And we in dreams behold the Hebrides.

If such sentimentality ever existed in Ness, however, it probably survived at the north end of the end. Even today, the sheiling lives on, in places like Cuisiadar and Biliscleit. Every summer a small handful of people still travel there for their summer holidays, venturing out along the peat road which leads from the most easterly villages in the district, Skigersta and Adabrock. There was even in the twenties a church and manse built on its clifftops to tend to the souls of those living there in brief exile from their crofts and homes. Identical in every way to its counterpart in the village of Lionel, it used to ring out hymns – like 'Bringing In the Sheaves' – across that treacherous landscape, songs forbidden on the other side of that community where only psalms could be sung. Crofters, too, were dipped and baptised by the preacher of that church, John Nicolson or 'Ain Fiosaich, plunging flesh and soul into a moorland loch.

Yet most of all, it was a place of wonders, one in which the people of Ness gained glimpses of another world in more than one sense. Some of the older generation of the district's men and women used to tell me that the first time they saw an orange was in the fingers of John's blind wife, Nora – an American woman and member of a rich and well-known family called the Cushings who had settled with her husband in these parts. Their eyes were round with astonishment as they watched it spin within her hand. It suggested a richness and fruitfulness glimpsed – for the first time – within a landscape that often appeared bleak and inhospitable, an exotic blaze of colour in an environment sometimes as dark and forbidding as Nora's gaze. Most of all, though, that fruit was a sign and symbol of the brave new world that existed across the narrow waters of the Minch, one that had such people as Nora Cushing within it.

'It was the first time I ever wanted to leave home,' one of them informed me as a young man. 'Knowing there was a different way of life like that out there.'

And there were other dreams, too, for the young men and women standing on that headland, one where you could behold not the confining limits of their native Hebrides, but instead vast stretches of the Scottish mainland. It was an edge and promontory where young people could view the outline of the Sutherland hills – the alien outcrops of such peaks as Suilven and Stac Pollaidh, as well as the vast moorlands rolling endlessly both beyond and around these mountains clear and vivid on the horizon. In its sheer scale, both in terms of height and depth, this 'frieze of mountains, filed on the blue air' of the north-west Highlands, as seen by Norman MacCaig in 'A Man In Assynt', dwarfed the low and narrow island on which it stood. The landscape was in no sense a prairie, not even a narrow one. Instead, it sported crags, foothills, ravines, gullies and glens, routes and channels that twisted and turned in the most unpredictable of ways. It suggested unknown roads and secret, hidden localities, ones in which adventures and mysteries might be found.

It made them think of other territories they could explore, not just spending their lives confined and trapped within its shores. Instead they could leave home and seek out locations as exotic as Newfoundland with its magical range of place names, each one pledging their own promise of delight.

Pleasantville. Fair Haven. Murray's Harbour. Heart's Desire.

CHAPTER TWO

Lyng (Danish) – Heather

A short stay on the island of Iona a few years ago allowed me to stumble, unprepared and unaware, into other people's worlds.

On one occasion, my entry was brought about by an unexpected downpour of rain. Looking for shelter, I raced into nearby St Oran's Chapel, just beside the famous abbey that stands at the centre of the island. It was a small building where I had been a short while before. This occasion was, however, different. Together with the swallows dipping in and out above the doors, I was met by the scent of incense, the chant of prayers. A man with a rich, sonorous English accent was standing before the altar at the chapel's end. He was dressed in the kind of clothing that might have made him a fashion icon when St Columba founded a monastery here in the sixth century. Long, grey hair knotted and plaited below a black hat, complete with flowing drapes, he possessed a snow-white beard that flapped almost as much as his dark robes. Around his neck, there were religious vestments and a large gold cross. It was easy to believe that I had slipped back in time, meeting a contemporary of St Columba – St Ciaran of Clonmacnoise, perhaps – who was still conducting worship in the islands.

Behind him an older woman, dressed in the dark wimple of a nun, sang the liturgy, part of the rites and rituals of their faith. She was accompanied by others who occasionally spoke or crossed themselves in response. They included a red-headed man with crutches who told me that his old spiritual home had been a certain football stadium found in the east end of Glasgow, the one often called Paradise by the team's green-and-white-striped supporters. Now – he said – he had found his own vision of Paradise through the faith and structures of the Orthodox Church, an institution based much further to the east. Afterwards, we walked together up the road to a retreat where the members of the faithful stayed together. I strolled onwards after that, heading to the hostel where I was going to rest a few nights.

It was there I met Fiona Coates, a tall fair-headed Australian botanist who was over in Scotland for a short time. The centre of her pilgrimage was not, however, Iona, like those in the Orthodox community I met, paying honour to some of the early saints of the Christian Church. She was here to do a little 'wild swimming', plunging through waters near where the writer George Orwell had almost drowned and at which even Roger Deakin, the advocate of that cause, had balked – the Corryvreckan whirlpool that lies in the narrow channel between Jura and the uninhabited isle of Scarba. We stood for a while together at the hostel window where we were staying, talking about the madness of both our lives – her pleasure at meeting the challenge of sea and deep water, my life travelling in circles around Scotland's edge trying to teach young people to see their world through new eyes, employing both Gaelic and English in the task. The stillness of the evening helped us relax from all this. Apart from Tràigh Bhàn nam Manach, the white strand of beach that according to its Gaelic name was once stepped on by monks, perhaps during St Columba's time all these years ago, there was little but sea and tussocks of heather around us. The redness of the setting sun made the nearby Isle of Mull glow as if it were lit by flame.

'It must be very strange to see all that peat around,' I said at one point.

'Not really . . .'

'Oh?'

'There is peat in Australia, you know?'

I didn't. My vision of that continent was that it was largely built on sand, dry, sunlit and often parched of water. Fiona, however, told me about areas like the Australian

Alps and the Gondwanan landscapes of Tasmania, in south-eastern Australia, where sphagnum moss and brown fibrous sedge peat and other characteristics of the moorland could be found.

'Not quite in the same quantities as here. Just in small patches in the valleys and plateaux of the mountains and moorlands. There are even coastal peats and tropical peats in Australia as well.'

'Oh.'

As Fiona talked, I became aware that for all their size, these areas were, in themselves, under threat. There have been occasional housing developments, especially tourist accommodation, encroaching on that landscape, remote as it was. The distance and the roughness of the terrain also challenged men in other ways. They might ride sometimes or drive off-road vehicles across their length and breadth – in much the same way as I have sometimes seen young men drive their quad bikes out into the moor, speeding across scuts of bog cotton, disturbing rabbits from heather-blazing hills, tearing tracks and trenches across great depths of peat, splashing through the edge of a loch or bog. More commonly, however, there were animals grazing – cattle or horses chomping all they could find within the landscape; the latter, to my mind, like the Shetland ponies one occasionally glimpses running wild either in the islands that are their place of origin or the Scottish Highlands as a whole. The ponies inspire a great deal of affection, both among locals and visitors. The attitude of Australians, however, to their equivalents in that continent seems a lot more mixed.

'We call them brumbies if we're feeling a little romantic and sentimental. Feral horses if we regard them as a pest.'

She went on to speak about the way her country decided to portray itself to the world in the opening ceremony of the Sydney Olympics in 2000. It chose 120 stockmen and women dressed in bush clothing and mounted on stock horses, all riding to the soundtrack of a movie called *The*

Man from Snowy River. Its storyline was based on a nineteenth-century poem by the bush poet Banjo Paterson, one that told of how:

> the bushmen love hard riding where the wild bush horses are,
> And the stockhorse snuffs the battle with delight.

Fiona smiled at this, making clear she believed in the opposite view, that their counterparts – feral horses or the 'wild bush horses' – were a massive nuisance, churning up and polluting bogs, spreading weeds, laying bare drier areas where they rolled around.

As for cattle and sheep, they had now largely been removed from National Parks over the last few years with the intention of what Fiona termed 'rehabilitating' high country landscapes, including bogs. This has sometimes been controversial. The decision to ban grazing in the Alpine National Park in Victoria in 2006 by the State Labor government was particularly contentious.

'There were some unsuccessful challenges following that decision after changes in the government,' Fiona declared. 'Alpine grazing, as it's called, once operated under a licensing system. Cattle would be taken up to the high-altitude valleys for summer by graziers, to the grassy snow plains and woodlands. Sheep might also accompany them. Families would stay in huts in the mountains before bringing the cattle down in early autumn, before the cold weather set in.'

'That's very much like our *àirighean*, or sheilings,' I said, nodding.

It was also believed that these areas were, just like our moorland, set on fire to encourage new grass growth for the following season. This undoubtedly depended on the individual, so that leaseholds were burnt at varying frequencies – some each year, some hardly at all and some occasionally. Once again, this was part of my experience too. Some crofters seemed to set small patches of land on fire annually; others hardly did it at all. Again, as in my native Hebrides, a lot of the huts where men stayed have also been destroyed or burnt down in the last few years. Fiona regretted this, believing that they are very interesting and, even though representative of environmental damage, have their own place in history and culture. In this, she was different from some conservationists who tended to be opposed to their restoration. She disagreed with them, believing that, rightly or wrongly, they represent an influential element of the history of that environment.

In Fiona's view, however, grazing by domestic stock was the most longstanding issue that has caused degradation of the high country. Cattle, sheep, horses, even sambar deer introduced from Asia near the end of the nineteenth century not only damaged the peat and structure of bogs but, through vegetation removal, caused an increase in surface-water run-off. Higher volumes of faster flowing water running downhill have scoured out channels and led to deeper and wider incised streams. As a result, more water is sucked in and the edges of the bogs become progressively drier. Added to this is lower annual rainfall owing to climate change, and increased fire frequency. Species that grow upslope in drier conditions, such as grasses, tend to encroach into the peat as the bog dries out. This in turn makes the peat more flammable once the wetland species retreat further towards the central, wetter areas.

And then there was man. He might come on foot, damaging the landscape. Despite the sheer quantity of Skyemen, Scotsmen and Irishmen who settled in the area, there is no history of peat being used for fuel in Australia. Instead, the new arrivals relied on wood, easier to obtain throughout much of their new continent – and certainly less back-breaking for those who remembered the labours of the old. The mountains were also visited annually by Aborigines, among the original users of the high country.

Together with vegetation and meat, they also sought insects like Bogong moths to supplement their diet.*

The modern menace might also be the camper, occasionally setting off a blaze through the landscape by not watching a campfire carefully enough. Sometimes, regrettably, there have been instances of arson, people setting tracts of land on fire for a multitude of reasons: for the excitement smoke and flame spark off in a dull and unhappy life; for the sight and sound of sirens clanging through the night; for revenge or profit; for no discernible reason at all. These are seldom started in the Alps but those that are lit can travel enormous distances and end up setting the mountains ablaze.

Yet it is not only this that has caused an increase in the quantity of fires in Australia's peatlands. While large landscape-scale fires at high altitudes have been relatively rare over the centuries, this has not been so in recent years. Together with a rise in the total of blazes caused by humans, fires have become increasingly common in south-eastern Australia in the past decade during periods of extended hot, dry weather, ignited even on occasion by lightning, or multiple lightning flashes during summer storms. Heat plays its role too. In a continent always prone to both flooding and drought, the slight rise in overall temperatures over the last while has not assisted the survival of this landscape.

Fiona pointed out that there are very few mountains that have not been burnt in bushfires in south-eastern Australia in the last 15 years. The exception was Mount Baw Baw in West Gippsland. Recently, the most tragic fires irreversibly damaged landscapes in northern Tasmania, killing centuries-old pines found nowhere else in the world and burning enormous tracts of vegetation unique to Tasmania, including bogs, many

* Nowadays, politicians are among those who might accidentally come to eat them. According to my research, the moths have been known to gather in the Australian Parliament in Canberra, swooping down into the politicians' open mouths.

of which have dried out in recent years. She described it as one of the biggest and most dramatic environmental disasters in recent times, especially since it has occurred in this changing climate. The high temperatures and lower rainfall that contributed to this will continue to hinder recovery to its pre-fire state, with some burnt vegetation predicted to be permanently replaced by plants adapted to drier conditions. Remarkably there was little public outcry about this, for all that the pines that are found there go back to the days when scientists believe there was one super-continent in the southern hemisphere. Known as Gondwana, this land mass comprised Antarctica, South America, Africa, the Australian and Indian subcontinents as well as the Arabian Peninsula – the last two having shifted north. Clearly, if the loss of ancient pine trees fails to alter people's attitudes to their landscape, there is an even smaller chance that the fate of peatland might influence their viewpoint. Not only is it less glamorous than pine trees, it is also an environment known to fewer people, particularly in Australia.

'Bogs are outside people's normal experience,' Fiona said. 'That's unless they are a bushwalker in Tasmania and they find themselves immersed in it – literally! The bogs of Tasmania are infamous, although a lot have been boardwalked now on the most popular walks.'

She continued to speak, telling me how alpine bogs and fens have legal protection under Commonwealth legislation. Despite this, most Australians are unaware of their existence or importance. In fact, they were far more likely to recoil at the entire notion of sticky wet peat in bogs full of snakes.

In her view, the single most precious thing about bogs, worldwide, is their ability to preserve pollen grains. Bogs contain layers and layers of environmental history dating back thousands of years. Palynologists identify these pollen grains and reconstruct past floras, and from that they can get at least some idea of past climates, and how plant composition changes with climate.

'It's how we know what the landscape may have looked like during the last Ice Age, compared to how it looks now,' she said. 'Peatlands not only keep the world's natural history safe, but can also give us a benchmark to develop a sense of how rapidly the world is changing now.'

My sense of the world swirled with the same power and intensity as the whirlpool Corryvreckan as I contemplated the conversation later. It made me see the thick slabs of peat oozing between my fingers as I worked on the moorlands of Lewis in an entirely new manner. In an odd way, they had become like these encyclopaedias I used to squeeze out of library shelves in my youth, big and bulky ledgers to hold, storing much of mankind's experience and knowledge in their weight. I became conscious, too, that for all that moor was no less than 7,000 square miles, or 20 per cent, of Scotland, it was much more of a universal landscape than that. It existed in Newfoundland or Nova Scotia on the far side of the Atlantic, as well as the mountains I glimpsed – Suilven, An Teallach, Ben More Coigach – as I looked across the Minch towards the mainland on a still, frost-laden day.

The whole world was now peat, something I had half-considered might be true on those wet and stormy days I was trapped on the island in my youth. Around 3 million kilometres, or 2 per cent of the total land area of the globe, it existed not only in my immediate vicinity, beyond house windows blurred by continual rainfall, nor just in the northern hemisphere, but at the southern end of Patagonia in Argentina, the Falkland Islands, throughout Indonesia, New Zealand, and even the Kerguelen or Desolation Islands, not far from Antarctica, described by songwriter Al Stewart as 'the loneliest place in the map'. It was even in Central Africa, as I discovered through a chance conversation with my friend Gordon Dargie in a Shetland supermarket. Excitedly he informed me that his daughter Greta had, in a team with Dr Simon Lewis from Leeds University, discovered a peat bog 'the size of England' in the Congo basin, one that contains billions of tons of the same substance – lying some 7 metres below the surface – with which I worked during much of my teenage years. Within an environment where they often encountered gorillas, elephants and dwarf crocodiles, it appeared that their greatest obstacle in charting the 40,000 to 80,000 square miles of peatland was the dampness of their feet, even in the short span of the year that they were able to walk there. Decaying at a faster rate than its counterparts in other, colder latitudes of the world, this type of landscape is rare in the wet and warm tropics, unlike the northern hemisphere, where it stretched across many hundreds of miles.

Despite the ubiquity of peat in my landscape, it was in so many ways a mystery to me. It was for this reason that I decided to change walking companions, to leave behind my dog and go out there instead with Jonathan Swale, an officer for Scottish Natural Heritage in Shetland. Originally from North Yorkshire, Jonathan is known for having discovered the size and scale of what is now regarded as the largest cave in Britain. At Eshaness in the northwest of Shetland, Calders Geo sea cave is over 18 metres tall and might be as much as 5,600 square metres altogether, much larger than the Frozen Deep cave found under Cheddar Gorge in North Somerset. Its other local name, Hoiden Hols, suggests its use in the past to escape press gangs or hide smuggled goods.

Recognised, too, for his sea-kayaking skills, Jonathan is extremely well-fitted for dealing with this kind of surface. A young woman, Emma Sandison, who is on work experience from the Anderson High School, the largest secondary school in Shetland, is also coming with us. Young, slim and fit, she keeps pace with Jonathan, stride by stride, as I watch. Having spent much of the winter indulging in that traditional northern pastime of sitting slumped in a fireside chair, I fear the outer limits of my fitness will be explored just as much as the moorland.

And so it proves to be. Yomping across the stretch of moor those in Shetland call the Black Gaet, I am conscious that perhaps the time is overdue for reviewing my winter routine. I walk, walk, walk, envying the way Emma leaps over streams and scrambles across ditches. I spend too much of my time watching the backs of Jonathan's yellow wellingtons, noting the initials 'JS' printed there. It is just as well they are kind to me: Jonathan making sure that a barb of a fence is freed from my jacket; Emma stretching out a hand when I take a peat slide. Jonathan is just as generous when explaining the nature of the landscape, blanket bog like much of the Ness (Barvas*) Moor near where I grew up. From time to time he bends over, picking up a small piece of damp moss in his fingers. He squeezes, showing the drizzle of water contained within.

'They can hold 20 times as much water as they weigh when they're dry,' he declares, noting that there are many as 12 species of sphagnum moss in this small area, the reason he always takes visitors here. 'There are 26 found in Shetland as a whole. There are as many as a 120 both here and elsewhere.'

He possesses a knowledge of their names that baffles me, reminding me of the occasions in my life when there were identical twins in my classroom (*Was that Christina or Anne who just did that?*) each time he picks one up. Looking at it briefly, he provides a Latin name, perhaps *Sphagnum magellicanicum*, with its red crown, or *Sphagnum papillosum* or *Sphagnum cuspidatum*, found only where the bogland is pristine and clear, or *Sphagnum fuscum*, where the heather grows through. Each one seemed to have its own purpose on this landscape, its own role to play. There is even one that looks like green scum on the edge of a pool. He squeezes it in his fingers before showing us what it resembles once it has been drained of water.

'It looks like a kitten that's been standing too long in the rain.'

* Barvas Moor is, of course, its proper name. For reasons totally connected with local patriotism and chauvinism, I will probably continue to use the terms 'Dell Moor' and 'Ness Moor'.

The capacity of sphagnum moss to absorb liquid has always proved extremely useful. In earlier centuries, it was used to line nappies, soaking up urine and excrement too when someone in a household was suffering dysentery or diarrhoea. Women employed it as a sanitary product; both genders as a way of treating haemorrhoids. In the modern age, it is still sometimes used to cleanse home pools and spas, occasionally used in public swimming pools. It is assisted in this by possessing sterilising, cleansing qualities. It even contained iodine, a product I can recall from my childhood, its dark purple colouring daubed on a cut knee or arm.

As a result of this, sphagnum moss becomes even more valuable at a time of war. Both in the First World War and during many conflicts beforehand, it was used to staunch blood and to create clean and sterile dressings for wounds. Women and children used to go out and collect it – either by hand or rake – on the moors of all these islands, delivering it afterwards to the nearest special 'sphagnum collecting centre'. From there, the 'peat moss' was sent to 'sphagnum dressings sub depots'. These existed throughout these islands from Dartmoor to Aberdeen,* the city of Cork to the West Highlands, places where the moss would be dried, treated and stuffed in muslin bags before being sent to the war front. It was not only because it was cheaper that it was more useful than scarce cotton wool. It could be used to stuff pillows to prop up what remained of the leg or arm of an amputee. Small pads were also employed to assist those who were in hospital suffering shell shock, their night-time terrors assuaged by their belief that they were suffering more of a physical than a psychological wound.

And then there were the more exotic uses. The Inuit used to pad out their boots with dried moss when they stepped into the ice and snow, keeping their feet warm and

* In Aberdeen, 22,000 dressings were produced each month. By the end of 1917, more than 300,000 per year were coming from Ireland.

dry. The Chinese employed it to cure eye conditions. An English firm, Peat Products (Sphagnol) Ltd, used to create a variety of products to show its benefits. These ranged from 'Sphagnol Toilet and Nursery Antiseptic Soap' to special shaving foam, cures for skin diseases and veterinary products. Operating from the end of the nineteenth century to the beginning of 1969, the company even advertised its ability to restore hair to an owner's precious pet dog, a chow-chow. Whatever its efficacy in that particular case, there is little doubt that it has the ability to perform slow miracles in the Shetland moor, making the bog grow.

In an entry for 1827 in the *Diary of An Irish Countryman*, the Irish writer and hedge-school master Amhlaoibh (Humphrey) O'Sullivan notes, simply and basically, what the ordinary people around him had already observed:

> *Chim cionnas fhasan an mhoin. Ta luibh air an ngairrean siad susan. Fasann so sna poill mhona agus d'eis feo do, iontaoin se ina ghar, agus d'eis iomad blian liontarr an poll mona mar so, agus bhunn ina phort aris.*
>
> I can see how the bog grows. There is a plant called 'susan' [sphagnum], which grows in bog holes. After it dies it changes into peat, and after many years the bog hole fills up in this way, and becomes dry bog again.

Jonathan's explanation is somewhat more full and fluent than that, though in essence it was much the same. I watch him crouch beside the edge of a peat bank, pointing out its layers, taking small pieces from it and rubbing them in his fingers. He notes its distinctive three levels, characteristic of the blanket bog I often saw in my youth. The first of these is the turf on the surface,* which my father removed with his spade, a tangle

* Confusingly sometimes the peat itself is referred to as turf. This is particularly the case in Ireland. This is a word which, pardon the pun, also has deep roots, coming from the medieval *turbarium*.

of heather roots, sphagnum moss and cotton grass, which flourish upon this acidic soil. There are a few reasons why the moor consists of this type of ground. The first is simple – that the organic matter and minerals that break down over time are acidic in nature, and this in turn makes an acidic soil. Another is the excessive rainfall that falls in these locations; one reason, for instance, why the west of both Ireland and Scotland have more bog and moorland than their eastern edges.

At one time, the turf on the moorland's surface was useful as an underlay for a thatched roof and even formed the foundation and walls of small buildings, the stalks of heather and cotton grass creating a tight mesh, impossible to shift or dislodge. Jonathan pulled at the cotton grass with his fingers to illustrate this, telling Emma and me that the colour of the stalks is why the moor takes on a russet shade in autumn, shading the entire sweep of the landscape. I recall too how our own moorland often looked as if it was sprinkled with snow: it was the bog cotton flagged with little clouds that enabled its seeds both to be kept wrapped and warm as well as to be blown away in the wind.

'It's a sedge rather than a grass, despite its name,' he says. 'Tough and hardy.'

The next spongy layer is called the acrotelm. This was where my father obtained his *mòine bhàn* or white peat, the light fibrous material that was laid upon a flame to keep it smouldering overnight.

It is peat that is still 'meeting its last definition', to use a phrase from the Seamus Heaney poem 'Bogland', decomposing in the immediate area below the upper layer of turf and able to soak up rain. Largely made up of the different, decomposing types of sphagnum moss, heather and other plants found in these parts, it has been described by Irish writers like Gerard Boate in the seventeenth century as the fuel taken out of the 'dry bogs' or 'red bogs'. It is said by him to be: 'light, spungy, of a reddish colour, kindleth easily, and burneth very clear, but doth not last'.

From my own memory, it also took a considerable length of time to dry – one reason why we threw it out into the heather, allowing the wind to swirl and swish around it. There were occasions when we stepped out in the moors weeks after it had been cut, finding it even damper and heavier than it had been before, soaking up rainwater like a sponge.

It will take many years for its 'last definition' to emerge, its transformation into the third layer, the catotelm or black peat. This cannot soak up water. In its undisturbed state it's permanently saturated, and when it dries almost waterproof. It is this that is the 'blue clod' beloved of Shetlanders, their motivation every summer for going to the 'hill', the term by which most natives of these islands refer to the moorland. It is the reason, too, why one of the favourite bands in Shetland – playing ska, of all musical genres – rejoices under the name of Pete Stack and the Rayburns,* gentle advocates of the group Madness's insistence on going *One Step Beyond*. The blue clod makes the cold of winter bearable, as celebrated in the words of the poet Vagaland, aka T.A. Robertson:

> your ain folk's wylcom warms your heart
> laek da haet o da god blue clods.

It is sphagnum moss that is behind much of this, the creator of a complex chemistry that provides us with this combustible fuel. Despite this, it is a very slow process. Feehan in his encyclopaedic work *The Bogs of Ireland* notes that it grows 'on average between 10 and 100cm every 1000 years' or '1mm every two years'. On occasion, however, it exceeds this. There are accounts of places such as Hanover in Germany where rates of increase as much as 4 to 6 feet (almost up to 2 metres) in a period of 30 years have been recorded. There are also locations in Ireland and elsewhere

* A reference to the Rayburn stove that was once present in every house in the Highlands and Islands.

where peat moss has accumulated around a metre in 20 years. Its depths shake and tremble as Jonathan, Emma and I walk across its acres, disturbing a flock of sheep grazing on thin pickings, sending a hare scurrying before our footsteps, being overseen, too, by a clutch of ravens that croak hoarsely across a domain that is partly under their kingship. I must admit my feet were made lighter by having been spared the fate I read about in Tim Robinson's book *Connemara: Listening to the Wind*. He quotes an extract from Robert Lloyd Praeger's *The Way That I Went*, where he tells of a clutch of botanists coming together on a peat bog in Roundstone, County Galway. In order to determine the nature of the peat there, he notes that one of the men, possibly Jessen, later described as 'the great archivist of the Irish peat bogs':

> 'would chew some of the mud brought up by a boring tool from the bottom of the bog, to test the presence or absence of gritty material in the vegetable mass. But out of such occasions does knowledge come, and I think that aqueous discussion has borne and will bear fruit.'

The blanket bog in the Black Gaet is not the only place where our feet grew wet that day. They squelched and squished, too, a little farther south, when we stepped across a section of the township of Cunningsburgh in the south mainland of Shetland. At first sight, this looked different from where we had gone before – much more rich and fertile, a meadowland for the community when men set out with scythes to sweep away the tall grass that grew there. I envisaged something like the landscape of my childhood where every small patch of cropped grass seemed to possess its own haystack and haycock, where I used to go out with either a heavy wooden rake or a hay fork in my hand, working alongside men like my father and uncle, and our neighbour Angus Graham.

Yet appearances can be deceptive. This was peatland too, formed out of wet and dampness. At one time, say 12,000 years ago, there might have been a small loch here, which became filled in by rush, tall grasses, sedge, choking the flow of water and transforming it into land. Jonathan points at the hills that surround three-quarters of this waterlogged soil. Nowadays they are occupied by crofthouses and their gardens. They look out onto a landscape we associate more with places like the Norfolk Broads, places where God possessed a clear view except on these days when mist rose from its surface, infecting 'her beauty' as *King Lear* described: 'You fen-sucked fogs drawn by the powerful sun, To fall and blister!'

At one time, further south in locations like the Norfolk Broads and the edge of the Netherlands, it was places like these where they dug out their winter fuel, squeezing it out of the darkness of the ground below. This was a fen, peat-forming wetland fed by both groundwater and the streams that trickle down to swamp low-lying land. It is different from the blanket bog we have encountered before. One of its few similarities lay in how the wheels of tractors can sometimes become trapped within its dampness. For all that it was largely peat, it was not acidic. Flowers flourished on its surface, proof that – for the botanist – there are riches here. Jonathan once again dips a knee to lift a clutch of flowers. There are many more to be found here than out on the blanket bog.

'Ragged Robin … Marsh Sunflower … Heath-spotted orchid … Self-heal …'

'Self-heal?'

'Yes, Self-heal.' He smiles. 'Marsh forget-me-not …'

I checked up when I arrived home, noting that the plant was used for all sorts of wounds and even heart disease at one time, in the days before World War Two. Its richness and fertility contrasts with the next stretch of moorland to which Jonathan accompanies me. Again in the south of Shetland, this is not far from the high ridge that exists throughout much of the island, confusingly known to the locals as the mainland. Running through the island like a bare spine, it is free of the same thick covering of heather and sphagnum moss that flourishes on lower ground, eroded partly by weather, the sweep of snow and storm. Jonathan takes Emma and me just above Channerwick Burn where in September 2003 apocalyptic peat slides took place, dark avalanches that toppled fences and barricades, blocking the road between Shetland's main town, Lerwick, and its airport in Sumburgh. Several cars narrowly missed being hit by the torrent of dark, impenetrable sludge as some 30 peat slides occurred in the area after several hours of torrential rainfall.

'It had been dry for quite some time before that,' Jonathan declares.

And when the rain came, it was not contained by the spongy acrotelm, which was inundated with more than its weight of moisture. Instead, within a short time the downpour saturated the dark catotelm (black peat) below. It was a situation not assisted by animals grazing, as Fiona Coates pointed out was the pattern in Australia. It was not helped either by the peat cutters who had innocently worked down below, cutting away their fuel for generations. The dark debris gushed downwards, as it had several times in Shetland's history. It had happened in the thirties, fifties, eighties and nineties. It has been happening more frequently in various locations throughout Shetland in recent decades – a Corryvreckan of peat with thick dark breakers, riptides of solid black at war with each other, headbutting and crashing one another in continual conflict as they plummeted down towards the stream below. When it was over, it left behind a strange and alien landscape in its wake. Boulder clay stripped of all that had covered it for centuries. Clumps of peat that jutted out like small standing stones. A world that looked as if it had pitched and shattered, smashing into a calamity of broken pieces.

And not just here but elsewhere. Over the last decade or two, this kind of flood has occurred in other portions of these islands. It occurred a month after Channerwick in the village of Derrybrien in County Galway. In October 2003, half a million tons of peat slipped downwards from the site where wind turbines were being constructed, killing 50,000 fish – from perch to brown trout – in its flood of blackness, a case that resulted in the European Court of Justice finding the Irish government guilty for its lack of proper oversight. It is a phenomenon seen in January 2016 in the towns and communities in proximity to the North Yorkshire moor: Hebden Bridge, Otley, Whitby, York, the black sludge thundering down from the peatland, engulfing man in its wake. In this, they are imitating an event that occurred in Chat Moss near Manchester in the sixteenth century, during the reign of Henry VIII. In Robert Chambers' *The Book of Days: A Miscellany of Popular Antiquities*, John Leland tells of how:

> Chat Moss* brast up within a mile of Mosley Haul, and destroied much grounde with mosse thereabout, and destroied much fresh-water fishche thereabout, first corrupting with stinkinge water Glasebrooke, and so Glasebrooke carried

*Chat Moss had an important role in history. At one time, its existence put at risk the prospect of the Liverpool–Manchester Railway until George Stephenson the engineer succeeded in creating a railway line across its expanse in 1829. The line was 'floated' on a bed mingling tree-branches and heather ropes – much like the ones Highlanders and Islanders used to employ to ensure that the thatch on their homes was held in place during storms. This was topped with tar and covered with rubble. After that, the Rocket travelled safely across Chat Moss, the first of many steam engines to do so across moorland. There is little doubt that – as can be seen in places like Rannoch Moor and North Yorkshire – rail travel allowed people to appreciate this kind of environment for the first time. Before that, both foot and horse hoof had an unpleasant habit of becoming bogged down in it.

> stinkinge water and mosse into Mersey water, and Mersey corrupted carried the roulling mosse, part to the shores of Wales, part to the isle of Man, and some unto Ireland.

Raised bogs – a rare phenomenon, rounded in shape and occurring in shallow basins or on flat, low-lying areas where poor drainage waterlogs the ground and slows down plant decay – are particularly vulnerable to this. Solway Moss on the English side of the border is the location of a battle that took place between Henry VIII of England and James V of Scotland in 1543. Very few on either side were killed but around 1,200 Scots were later held prisoner, an event that – it is said – caused the death of the Scottish king through heartbreak.

More heartbreak and destruction followed when the 1,600 acres of Solway Moss suffered a bog burst on the night of 16 November 1771, flooding 28 nearby farms to a depth of 30 feet (9 metres). Fortunately no one was killed, with the farmers possessing the time and foresight to have noticed the movement of land. (There had been an event some two years prior to this.) Together with peaks and pyramids of peat, like small islands in the darkness of water, were human bones, silver coins, pots and ancient weapons, the remnants of great trees.

CHAPTER THREE

Heide (Dutch) – Heather

According to a legend of St Ciaran – reputedly the founder of Clonmacnoise, the monastery at the centre of Ireland in County Offaly – a vision inspired him to build his church there. It was of a great and fruitful tree that stood beside a stream at the heart of the country, its branches sheltering the entire island. Birds came to gather the fruit that clustered among the leaves, taking it in their beaks across the seas to other nations. Some, too, floated in the waves at its base, taken by the roll of tides to other fragments of the world. When he spoke to his friend St Enda about this dream that had disturbed his sleep, the other monk interpreted it for him. 'Go from the Aran Islands,' he said, referring to the place where they were both staying, off the western edge of Ireland. 'You are the great tree from whom many will receive blessings. All who live in this country will be nourished and blessed by your prayers. Build a church in its centre, one that stands by a stream.'

And from these words came Clonmacnoise, a cluster of ecclesiastical buildings and other structures that date from around the ninth century, a short distance from the small town of Shannonbridge. One finds the site by following a loop and curl of roads that might have been engineered by St Ciaran himself, a cattle herder apparently in his youth. They sweep and swirl round fields and isolated houses, designed, perhaps, to resemble long bridges constructed over water; the moor like an ocean, the occasional ridge of rush, grass and heather like waves. There are, too, constant diversions, crossroads which offer individuals meaningless choices between one empty stretch of moor and another. For someone like me, used to the straightforward geography of islands where there are often only two directions to choose from, either to or from the only town, they are a network of chaos and confusion. I find myself missing signposts, overlooking the route, perhaps to Ferbane or Creggan, becoming lost.

Yet when I eventually reach Clonmacnoise, there is little doubt that my bemusement has been worthwhile. Of all the sacred destinations I have encountered in Ireland and elsewhere, it is one of the most impressive. Outside the walls of the site, there are the ruins of a thirteenth-century castle. It staggers and totters, looking as if it is about to tumble in on itself at any moment, as if it were willing its walls to level flat and become mere stones embedded in the earth once again. Within the site boundaries, there is a crumbling cathedral, a scattering of churches – some in ruins and others intact – and an array of high crosses. A replica of one of them, the Cross of the Scriptures, stands outside. The original is among a few now contained in the modern interpretative centre. The panels upon it tell of the crucifixion, rough figures hewn into the rock portraying the agony and glory of that event in what appears to our eyes crude and basic, but what must have appeared at that time marvellous and miraculous.

It is the two round towers, named O'Rourke's Tower and McCarthy's Tower, that most fascinate me. I spent a while contemplating what they must have looked like when monks crammed into their walls, keeping a weather eye out for any Norsemen sweeping down in their longboats along the Irish coastline, seeking a gold-ornamented crozier or cross. I picture them squashing through doorways halfway up these structures, their cassocks flapping as they scale wobbling ladders, kept upright as much by prayer as nails and rough wood. I imagine, too, these priests looking out across the landscape, much of which they themselves must have denuded of trees, building the roofs and sometimes walls of structures they prayed and worshipped within.

Or it might have been that day when, according to one of Heaney's poems, a ship appeared while they were gathered to pray together, astonishing them as it flew above the oratory. The view they would now have experienced would have been 'out of the marvellous' in another way: each car bringing modern pilgrims to their doors, emerging from these carriages in strange attire. All would have changed completely from the way it was seen back then. Even the River Shannon, a short distance away with its golden borders of rush, would have followed a different course. As for the rest …

It was Antony, one of the guides at Clonmacnoise, who provided the best testimony of the changes that had occurred. A quiet, thoughtful individual, he had recently returned to the area, having spent a few years working in a similar role at Kilmainham Gaol, the prison where some of the leaders of Ireland's Easter Rising were shot. A few days short of the centenary of that event, integral to Ireland's history as an independent state, he appeared grateful for the arrival of visitors in Clonmacnoise, each individual bringing the realisation that the long, slow, winter season was coming to an end. He spoke to me for a time about the change in both his job and location. For a while, he still travelled in from his old house in Dublin to go to work, navigating the long tangling network of roads with an ease which – after only a few days in the locality – I envied. He spoke, too, of the home to which he was returning to stay in a short time. It stood nearby in Bloomhill, a small, fish-shaped hill of green land which he described as having been surrounded by a sea of peat during his earlier life.

'With every year that passes, that tide is receding, the peat rolling back.'

To me as an islander with a swelling immensity of that same substance stretching out from my front door, it was immediately obvious to what he was referring. With every year they cut peat in our village, people had to go out a little further into the moor, the banks nearest the houses having been the first ones to be used for home fires. Some of the land left behind in the wake of all this hard work was solid and unyielding. Consisting of gravel and rock with an overlay of either short, cropped grass or heather, it lacked the softness and fluidity of the wide open spaces farther away where peat had not yet been dug. Below my footsteps, it felt like shoreline, solid and reliable earth, defined in a harder, sharper way than the more liquid quality of the moorland man had not yet visited with the cutting edge of a *tairsgeir* (a special peat blade).

This was particularly true during the early years of the development of the peat scheme in the Irish Midlands. One side effect – especially in these early days – was in how the drainage of peatland assisted local farmers, specifically in terms of the marginal land at the edge of their property. After the arrival of Bord na Móna, the Irish peat organisation, some of their acres were less damp and more fruitful than they had been before.

There was more to it even than this. What was not so immediately apparent, until I drove across this landscape, was how appropriate Antony's words were. The emptiness and flatness of the peatlands here resembled the sea with the tide having gone out, one that was denuded and devoid of life. Few birds sang above its desolation. No sprigs of heather bloomed. Sometimes gorse, or furze, marked its boundaries, spiking through a grey or blue sky. Its little yellow blooms seemed like signals of distress, a petty act of rebellion against all that mankind had done to the landscape. He had rendered it monochromatic in much the same way as mist sometimes does, blurring coastline and wave. The only real difference was in the singularity of the shade that had settled on the land here. All was the dull, khaki brown of a First World War battlefield. There was swell upon swell of soil. As if in the manner of open-cast mining, trenches had been dug at irregular intervals across its empty acres, scarring it for mile after mile. Occasionally black plastic bags, folded over these ridges, flapped in the breeze. This covered yet more milled peat, looking both desiccated and powder-puffed, harvested mechanically by the various machines employed for this purpose by Bord na Móna. Across this barren landscape, a narrow-gauge railway of various lengths and breadths ran. Its iron rails twisted through the centre of Ireland, throughout counties like Athlone, Laois and Kildare. Some were around 3 feet (915mm) in breadth, others a narrower 2 feet (610mm). Empty of engines and wagons, the tracks lay there with the wind sweeping back and forth, like those that ran across the empty tundra in the film *Dr Zhivago*, long trails that took the troops of the Red or White Army into battle with each other. Some looked temporary, spread across plastic matting and ready to be lifted and placed elsewhere once this particular seam of peat had been stripped and mined.

In some ways, this could have been the future of my native Lewis. When the soap baron Lord Leverhulme took over the island in the aftermath of the First World War, he had a vision of transformation of huge tracts of moorland into an area where agriculture would flourish, providing vegetables and soft fruit for a population that largely fished for the 'silver darlings' (herring) that swam around its borders. To begin with, this hinterland would have to be cleared of peat, sold commercially as fuel both home and abroad. After that, there would have been 'berries and brambles galore', as a geography teacher in my old school who wandered endlessly off topic used to tell us, claiming that our home town of Stornoway might back then have enjoyed the prospect of being like the city of Dundee, famous for its jam. Little of this sepia-shaded dream was, however, practical or likely to be realised. Partly due to events in Germany and Russia, the herring industry collapsed after the conflict. As for the expected fields of berries, Leverhulme had not taken into account either the poverty of the soil or the potent mix of wind and salt that whistled across the island, blighting the sweetness of any fruit that might have been planted there. There would have been years, too, when there would have been insufficient sunshine or dry winds to enable peat to be made fit for any household fire.

Instead of berry fields or market gardens and open fields of cultivated peat, it seems to me that, if Leverhulme's vision had triumphed, the expanse of moorland familiar to me in my youth could easily have turned out like Lough Boora Nature reserve, which I visited on my travels. There were some trees, downy birch, willow, a rare rowan or two, but it mainly consisted of grassland meadows, a scattering of flowers. Its length and stretch provided proof of the Irish poet Louis MacNeice's words, written in a prologue to an unfinished work: 'No more than peat can turn again to forest.'

This area was 'restored' by Bord na Móna after the peat had been worked and exhausted. Swans and ducks swam on a stretch of water. Anglers lashed the surface of a loch with their lines. Not only nature but art thrived. A sculpture with the title *Turf, Wind and Fire* stood on a shoreline, a symbolic beacon in the midst of that landscape, glistening like the soul band whose name it (deliberately) recalled. There was the Burrow Shelter, made from old tipplers at Ferbane Power Station, which was decommissioned at the turn of the twentieth century. These were employed at that time to clear the wagons of peat when they arrived in from the bog, rolling over and emptying them of fuel that would be placed on a conveyor belt and brought to the station. Another tippler resembles the Nissen huts I recalled from home, products of wartime in the islands, which allowed the visitor to view the landscape around them in a new and original fashion, tailgated at night by a speckle of stars. Most unusual of all was the Sky Train, an engine yoked to open creel-style carriages that – when the sun sank to dust level – would curve like the arc of a ghostly rainbow, freeze-framed in position as daylight ebbed away.

In his dreams of developing the industrial potential of the moor, Leverhulme had conjured up his own elusive rainbow, one that even came in the form of a railway crisscrossing the island. A few short lengths of it still remain within the borders of Lewis. If I close my eyes for a brief time, I can see, Dr Who-like, an alternative future for the Hebrides in these temporary lines still taking fuel to two of the remaining Irish peat-powered electricity stations, either in Kinnegad to the east or Shannonbridge to the west, linked to some six to eight boglands.* (Another, at Ferbane, not far away from Shannonbridge, was closed in 2001.) There were other lines that looked stronger, firmer, more permanent, designed to last like the length of Grand Canal they sometimes ran alongside. I stopped at various places where all three means of travelling the breadth of this country came together, rail, water and tarmac, crisscrossing one another. Sometimes, I could see railway barriers across roads, beside the Grand Canal that had once taken cargoes of fuel across Ireland to Dublin in the east. Displaying danger signs and red triangles, they creaked and whistled in the wind, rust flecking their surface. For all that there were 12 trains or *rakes* (a locomotive plus 16 wagons) in daily use 16 hours a day at West Offaly in April 2009, it would have been easy to gain the impression that they were not opened very often, the open road used more than rail.

As if to confirm this view, a lorry whizzed by on its way from a factory bearing a load of briquettes made from compact, shredded peat while I stood on a bridge to one township. Another truck passed slowly, hauling a cargo of tree trunks from some aspect of the country. It occurred to

* There were plans to introduce a peat-powered electricity station to the north of Scotland much later than the 1920s and to the Leverhulme era. Among other locations, such a development was considered for both Yell (in Shetland) and the Isle of Lewis as late as the seventies. A bullet dodged; a smoke cloud dissipated.

me that while the waters of the canal and the iron of the rail track now seemed empty, there had been a time when they had vied for the main mode of transport in these parts. The Irish poet Patrick Kavanagh had, for instance, sat on a bench beside the Grand Canal in Dublin and seen how:

a barge comes bringing from Athy
And other far-flung towns mythologies

Many of these vessels would have come from the peatlands near or around Clonmacnoise, a different way of bringing truth to St Ciaran's vision, the entire country being nourished and warmed not by the dead man's prayers and piety but from the depth of peat that lay at the heart of this country. Central to this were the peat power stations that generated electricity for much of the nation. Over the years, travelling through Ireland, I have passed a number, barely aware of them, only noticing, for instance, the one that stood near the town of Ballina over 20 years ago. It stood out in the bare landscape of the west of Ireland, a form of industrial development odd and alien among countryside that seemed in so many ways similar to the one where I grew up. I looked out the car window, noted the differences between these buildings and a home landscape where only a concrete water tower seemed modern and forbidding, and moved on. Until that moment, I had been unaware of even the possibility of using peat to provide electricity. If I had thought of it at all, I would have presumed that Ireland was like the Highlands of Scotland, the grains of light one could glimpse through the windows of isolated houses being sparked by the great concrete dams that provided hydropower.

It was a little different from that. Unlike the clean structures of many of the hydro dams I had seen over the decades in places like Cluanie in the north-west Highlands or Pitlochry in Perthshire, there was something of the fifties, the legacy of the Cold War about the structures I found in both Kinnegad and Shannonbridge. Reminders of the old Soviet Union were present in the striped barriers, rusting buildings, the leaden skies of the day I drove towards them. This was particularly the case with the latter, built a short distance downriver from the town that provided its name. Many years ago, in the Napoleonic era, the community had been the site of military fortifications constructed by the British. For all that the town had escaped this inheritance, the same could not be said for the power station. A wire fence surrounded it. Notices reminded people that entry was forbidden unless they had prior permission. Its tall chimney shuffled steam; its steel walls were blotched with rust. An ominous clank of mechanical noises seeped from within, as if some industrial version of 'Hey ho, hey ho' was being broadcast. Chutes delivered the milled peat into its walls, unloaded from the empty railway containers that rusted outside, lining up for a variety of purposes, bringing fuel to the station, taking ash away. The alchemy of transforming milled peat into electricity seemed a mighty, miraculous process, one that has to come to an end in 2030 as an important element in Ireland's attempts to lessen its dependence on this particular form of fossil fuel. In the meantime, the numbers involved in that work are dwindling, with only around 40 locals now found labouring in the power station. At one stage, as Antony told me in the visitors' centre in Clonmacnoise, someone in every house in the community used to be employed at a development that first began operating in 1965, generating 40 megawatts of electricity at that time. A second 40-megawatt set was built some four years later; a third of similar voltage came into being in the early seventies. After its arrival, it was not only lights that sparkled in the community. Silver began appearing in the pockets of those who lived in the townships nearby. Folded notes were found in wage-packets, enough for people to make a few fossil-fuelled journeys in their own cars, to purchase some of the small luxuries of the time. For the majority of the population, whose only previous options were a life of poverty or emigration, peat brought much more than heat and a homely scent. There was the sense that the power-lines threading out from Shannonbridge were creating a web of prosperity, one that bound those working in a nearby peat briquette factory, even youngsters studying at a school desk, within its narrative, keeping them both healthy and at home.

The last 60 years or so of the last century brought about these changes, though there is no doubt that the entire industry had deep foundations in the country for generations. Peat was, of course, burned in Ireland many centuries before that time. There are references to turf or peat in the *Senchus Mor,* the early law tracts, making clear it was being used as fuel as far back as the eighth century. Peat cutting was considered a share of the rent paid by ordinary people to landlords who possessed bogland on their estates from as early as the fourteenth century, acting as fuel for the household fires of rich and poor alike. This dependence on peat increased as woodland shrank over the centuries that followed, partly because wood charcoal was increasingly used in ironworks. According to some commentators, this 'commonly putrid and stinking' wasteland that created the fuel was caused by 'want of industry' and 'careless' Irish people who let 'daily more and more of their good land grow boggy'.* (The writer Gerard Boage contrasted this with the hard work of English settlers who drained their land, ignoring the fact that the latter group had the advantage of owning it, unlike the Irish who used it as commonage.) By the end of the seventeenth century, it became the most

* The most graphic example of this is the Ceide fields in North Mayo, where farming is said to have begun around 3000 BC but was over the centuries engulfed by blanket bog. (The date has recently been disputed.)

common fuel for household fires. By 1926, over six million tons of peat was being hand cut for people's homes, largely in the same slow, patient manner as my ancestors performed the task in the Hebrides. There are similarities between the lives of the rural poor that transcend all boundaries, the same form of labour bowing backs from Finland to the Outer Hebrides, Russia to County Kerry.

For all that, there was innovation in the way peat was used. Not far from Clonmacnoise, and now part of a nature reserve near the township of Pullough, there was once a peat moss factory in Turraun Bog that provided bedding and stable litter for the British artillery and cavalry regiments in townships like Birr. It was owned by the Farrelly family from the mid-nineteenth century until 1903, when it was flooded and sold to others. Following the example of similar enterprises in Germany, the Netherlands and Scotland, the peat moss litter it produced was also used for cattle, pigs and the packing of fruit and vegetables, particularly mushrooms. While dried peat is still sometimes used both in these ways and to absorb excrement from cattle wintered indoors, there is little doubt that with the beginning of the twentieth century and the arrival of the motor car and electric tram, this type of business did not generate the sort of income it did before.

There was also turf charcoal, which was used for smelting iron in Ireland from around the seventeenth century. Though it provided more heat than peat itself, its effectivity is open to question. One of my own ancestors fled from Dornoch in Sutherland on the east coast of Scotland to the Isle of Lewis around this time in the company of a bowyer (maker of bows) called William Murray, a man who became known as the *Gobha Gorm* (the blue smith) in the community.* William set up his smithy on the banks of the Dell river in

* The rather dramatic circumstances of this escape are told in my book *The Guga Hunters* (Birlinn, 2008).

my home village, fashioning bows for all those who wished to flex and wield them in the district. His arrival, however, kindled a great deal of hostility in the community, particularly among those who shared his trade. The reason? The quality of his work was so much better than theirs, his weapons rarely breaking under pressure. For this reason, he was accused of being a magician, an outsider who had come with dark and secret arts from the mainland. These accusations brought into being a darkness of their own. Gunpowder was placed by a rival in the heat of his forge, causing a small explosion that scorched his cheek blue-black, the origin of his nickname. The reason for the standard of his work, however, had nothing to do with magic – apart, that is, from the fuel he used. He imported coal from the only seam that existed in the Highlands. This was in Brora, near his home community of Dornoch. He would smuggle this into his smithy, telling no one about its arrival, employing it instead of turf charcoal in his forge.

Coal was not an option available to the Irish. It could not be found within the nation's boundaries. Instead, groups like the Irish Amelioration Society tried in 1850 to create the country's first peat station in Derrymullen in County Kildare, one that created peat charcoal for smelting and also to purify and remove the colouring from food or water. There were also various schemes to hasten the drying time of the fuel. Sheds were used. So were specially built drying chambers and electricity. In 1844, there were attempts in Kilcock in County Kildare to remove water from pulped peat by hydraulic presses. Many of these processes proved unprofitable. There was one occasion, for instance, when an ironworks was set up in 1851 in Creevelea, a place near Lough Gill in County Leitrim, where I stopped briefly on my travels. This endeavour ended when they discovered that it took 4 tons of wet peat to create 1 ton of peat tiles. After this, they required around 3 tons of these tiles to produce 1 ton of charcoal. The arithmetic behind this did not make much sense.

There were, too, in common with my own native Isle of Lewis, various attempts to create peat chemical works. As I was told by my former colleague Dr Ali Whiteford many years ago, the peat provided for these facilities was cut, dried and then distilled; the latter done by heat but without air. Four substances emerged from this process: a thick tar, a watery liquid, an inflammable gas, and a solid coke-like residue. These were employed in various ways. The tar, for instance, was the basis of a form of creosote daubed on railway sleepers to protect the wood; the liquid was used as a form of paraffin* and as wax for candles. Even during the late nineteenth century, this development was not seen as environmentally friendly. The liquid managed to poison the salmon in a nearby river. The gases – blown by the prevailing winds – frequently choked the citizens of Stornoway and caused many employed by the chemical factory to become ill. At one time, the building was shaken by a large explosion and much of its structure destroyed. Finally, it made the mistake of employing an individual called James McFadyen as its chemist. As well as polluting the foreshore near Stornoway, he misappropriated funds to create his alternative (and failed) venture, trying to extract dyes from peat. By the time he departed in disgrace for America in 1869, there was little future left for the Lewis Chemical Factory, a tale – as Dr Whiteford describes it – of 'comedy, farce and pathos, peopled by various scoundrels, buffoons, gentry, academics and a Hebridean hero or two'.

Yet despite all this, as I know from my own experience, for a community to possess peat was a boon to its existence. It might even be sold commercially, hoarded in the hulls of

* This method was initially developed by the Irish Peat Company in the 1840s. The story of the Lewis Chemical Works – which lasted from 1857 to 1975 – came from a leaflet written by Dr Ali Whiteford and funded by the Stornoway Trust and the Stornoway Amenity Trust.

Galway hookers heading out to places like the Aran Islands and the Burren where limestone crusted the surface of the earth, or even sold to those in nearby towns in creels or sacks. The alternatives to the fuel, when any were available, produced much less heat and light. They might dwindle away like those on the distant isles of St Kilda, who ran out of peat in the last years of the community. The fireplaces of the houses in Main Street lie empty today, cold and filled with stones which recall the date the settlers left for the final time. Or they might have been like the Aran Islands or the Scottish isle of Tiree, where some of my ancestors came from. It was often cattle dung that was burnt there, scraped with a spade from the shoreline or a field where the family cow had grazed. My friend Duncan Grant, approaching his mid-eighties and employed at the local history centre on the island, recalled being sent out in his boyhood years to find *sgàindeach* (cowpats) to put on the fire.

'You'd have to turn them over with a poker and leave them to dry. Then you'd go back and collect them for the fireside a while later.' He chuckled. 'There wasn't much heat in them. Not like coal or peat. Goodness knows how they got any warmth in the days they had only them to put on the fire.'

Duncan undertook this task because they possessed no peat on the island from around the middle of the nineteenth century. From that time onwards, they imported much of their peat from the Ross of Mull on the nearby Isle of Mull, sending a party of men out to labour there for some of the summer months. Many of the women would travel there too, living in outhouses on the island while they went out daily to turn over peats or lift them to set them to dry. There were years, too, when their peats lay mouldering where they were cut, unable to be used with the onslaught of wet weather, constant downpours of rain. Or there was the occasion when the Tiree men returned to find an old woman had stolen their fuel. They intercepted her on the way home, her back bent with her ill-gotten gains. According to Donald Morrison, a man from Bunessan in the Isle of Mull, the practice only stopped when a man from Tiree drowned going either to or from the peat cutting, falling into a storm-tossed sea near the distinctively shaped Bac Mhor (Dutchman's Cap), a rock of volcanic rock that forms part of the Treshinish Islands. After that year, the spades and *tairsgeirs* of the Tiree men were laid aside.

As mankind has often discovered about the price of coal, with men crushed below the weight of its density and darkness, the cost of even the best fuel can sometimes be too high.

PART TWO

Mòine Bhàn – White Peat

STORYTELLING

We raised from the moor
dark slabs of peat
lifted from the depths
with soiled hands and feet
firm upon the peat blade
before laying each to dry
flat upon the heather,
allowing wind and sky
to perform their part
in our labours;
placing them in stacks
beside which neighbours
gathered to assist us
in taking fuel back home.
Later, we'd scatter scraps of turf
on fires these nights we sat alone
listening to the tales they told
as that glow whispered of the days
when men walked far on moorland,
stories sparking into legends flaming in each blaze.

CHAPTER FOUR

Vaalea Rahkasammalturve (Finnish) – White Peat

Naming these things is the love-act and its pledge

Patrick Kavanagh

When I was younger, I spent much of my years in the company of a peat blade.

Each summer I would head out to the Dell moor with some other members of my family – my father, brother and uncle, perhaps – to cut fuel for the household's fires. It was a process that took a succession of journeys, making our way along a stony track that stretched along the fence of the village common grazings. On our first excursion, we would bring out a couple of normal domestic spades, wielding them in a variety of ways. Noting how the tractor that had carried last year's load home had breached the surface of the moor in certain spaces, we would set out to mend the damage that had been done, laying heather patches where the ground had been ruptured or torn by the whirl of giant wheels. Sometimes, too, Dad would ask me to look out for places where the earth stirred and wobbled, trying to ensure it could be made secure and firm. There were times when, as a young lad, I could recall bouncing across a stretch of the moor like this, impersonating Neil Armstrong or some other astronaut as he made his way across a similar but distant wilderness, one I had glimpsed through the crackling pictures on our black-and-white television. I would plant my spade in the barren earth just as I had seen him do with his spangled flag, clambering, perhaps, over an overhang of peat – a feature of this landscape that I now know is called a peat hag.

'One small step for man,' I might drawl.

And then another scrap of green or purple might be laid where a dark pool of water had formed, healing any break in our portion of the moor. If that were not done, there was the possibility of another journey taking place, one towards the centre of the Earth as a tractor skidded downwards, spluttering dark mud as it travelled, turning its wheels, spinning into eternity or whatever lies deep below. After that, he would try to put right a bridge or two where they crossed a moorland stream, placing further layers of turf down where its walls seemed likely to give way, adding stones, too, where this was deemed necessary. This would brace and strengthen it before a tractor rolled across, its trailer stacked with a tower of peats for the household's fires.

It was only when he was satisfied with these tasks of civil engineering that we would begin the true preparation for the work of the day. Using our spades we would remove or 'shave' the turf or *scraw* (to use the Irish term) from the edge of the peat banks, digging through the layer of heather a couple of feet from the verge, sliding blades underneath to cut away the roots that knotted it tightly to what lay below. Underneath this, the peat was finally laid bare – a moist, oatmeal shade compared to the dark, cracked crust that marked the boundary of last year's workings. This scab – one that had formed across the breast of the peat bank – looked as if it had been wrinkled and ravaged by the storms that often blew across the island, dried, too, by the occasional heat of the summer sun. We would stand beside its heights, some four foot or so, as we lifted the newly cut turves from the periphery of the peat bank, giving a small twist as we laid them down beside the bandages of heather we had put in place the previous year. This would make sure the heather 'fitted' with its neighbour, allowing the water that ran along the bank to flow freely in the direction of the stream nearby, permitting the landscape to remain unchanged even after the peat had been dug and removed. It would also provide a level stage of around 15 inches (40cm) or more from which to work when the cutting began. As a result of all this, the contours of the moor would remain the same. It was only its depths that had been altered; the distance between the heather at the top and the rock down below diminished and shrunk by our labours.*

And then the spades were set aside and another implement – the *tairsgeir* (the peat blade) – lifted. Cruder than its store-bought equivalent, it could be simply translated into English as a peat blade. Across Shetland, Scotland and the Faroe Islands, it is known by a variety of other names, such as *towskjairi* around Torshavn, *tushkar* or *tuskar* elsewhere. (The last of these titles is also given to a variety of peat-smoked whisky, distilled in Aberdeenshire, while *tushkar* is a type of stout, black as the substance the blade heaves out, brewed in Shetland. All of them are derived from the Norse term *turf-sgeir* or 'turf-spade'.) There are other difference, too, between one area and another. In the English fenlands, the implement was called a 'beckett' and shaped somewhat like a cricket bat. In other areas, it was more like a traditional spade. In Ireland, the cutting blade is narrow in a similar way to Shetland; in the larger community, however, it is called a *sleaghan* or slane. Only the strength of hand and arm is required to force it into the earth. A one-man team operates it. Tool and peat is lifted up and the peat twisted and turned to be placed high upon the heather-covered ridge to dry. In Lewis, the blade is much longer, a foot or so. This is because the peat is softer and deeper for the most part in that island. In many banks the long blade succeeds in cutting through the moist fuel like butter. As the peat was larger and heavier than its equivalent elsewhere, two people were required to

* In Ireland, the words 'turf' and 'peat' tend to be employed interchangeably. In Scotland, the word 'turf' refers exclusively to the heather-covered segment which lies on the surface of the moor.

complete the operation: generally the male on the blade, standing on the ridge that was called the *carcair* in my youth in Lewis, and the woman lifting. There was also a foot-sized step on the *tairsgeir* in order to ease the peat out of the moor, lessening the strain on an individual's arm.*

It is an activity that Dr Johnson takes notes of in his *Journey to the Western Isles of Scotland* in 1773 when he remarks that:

> The only fewel of the islands is peat. Their wood is all consumed, and coal they have not yet found. Peat is dug out of the marshes, from the depth of one foot to that of six. That is accounted the best which is nearest to the surface. It appears to be a mass of black earth held together by vegetable fibres.

There was even a hierarchy among different kinds of peat, which the good doctor gets wrong in claiming the best is near the surface. The first layer consisted of the *barr-fhad*, the first peat. It tended to be the one that contained the greatest network of roots and required a fair amount of pressure from a man's heel to cut through. This was even more true of the *cor-fhad*, or peat that was on the edge. Sometimes this was the one that was used to keep the fire going at night, used to bank or preserve the flame. At one time, people

* This was not – by any means – the first method employed to harvest peat. Pliny the Roman naturalist describes a technique now called maceration in his work, having witnessed the process in the first century near Bremen in northwest Germany. This involved the pulping and forming of wet peat by either hands or feet and leaving it to dry. It was traditionally used when the bogs were either extremely wet or under water. Widely used in the Netherlands, Ireland and elsewhere, it had better drying qualities and better calorific value than hand-cut peat. Similar to peat briquettes, it was lighter and easier to transport than those I heaved out of the bog.

placed three of these on the embers, accompanying the ceremony with solemn words of prayer, as recalled in Carmichael, *Carmina Gadelica*.

SMALADH AN TEINE	SMOORING THE FIRE
An Tri numh	The sacred Three
A chumhnadh,	To save,
A chomhnadh,	To shield,
A chomraig	To surround
An tula,	The hearth,
An taighe,	The house,
An teaghlaich,	The household,
An oidhche,	This eve,
An nochd,	This night,
O! an oidhche,	Oh! this eve,
An nochd,	This night,
Agus gach oidhche,	And every night,
Gach aon oidhche.	Each single night.
Amen.	Amen.'

A ritual that occurred the following morning is recalled by John Montague in his poem 'The Leaping Fire' where, in an elegy to a relative, he writes of how:

> I saw a miracle
> as you sifted the smoored ashes
> to blow
> a fire's sleeping remains
> back to life, holding the burning brands
> of turf, between work-hardened hands.

There was little sign of sanctimony, however, in the way the top covering of peats was treated. Keeping to a steady rhythm, they were flung as far as they could go. The outer edge of a line of peat rectangles that would later constitute a stretch of five or so would be cast out on the heather, its flight coming to a halt a couple of feet from the edge of the bank. Most of the time, this was the least valued peat for the home fire. Despite Dr Johnson's assertion, the *mòine bhàn*, or 'white' or 'fair' peat as it was called in Lewis, was suitable only for smoke and rarely for flame.

The peat in the second layer – 'spit' as it was known in parts of England – was slightly darker, more speckled than pale and gave off a degree more heat than the one on the crest of the bank. It was called either *an dàrna fàd* (second peat) or *fàd a' ghàrraidh* (wall peat). It was given this second name for a good reason; these rectangles were layered in a wall with their edges around 2 inches (5cm) from the side of the peat bank. The narrow space between this barrier and the rim of the bank was given a poetic name in my district – *rathad an isein*, or the birds' road or path – creating the image of a robin or a blackbird dancing along a thin margin of bog-brown earth. Despite the lyricism of this title, this slender gap had a prosaic purpose. The little bit of space it provided between the wall and the border of the ridge prevented those who worked in the peats from dirtying their clothes or shoulders from contact with the fuel. The wall, too, had a further reason for existence. Not only did it allow men to stack a ridge of peats, employing less space than laying them all flat, but the persistent wind would also whistle through the tiny gaps between each segment, circulating the air and permitting the fuel to dry more quickly.

It was probably at this level that I began to show my more demonic qualities in the peat bank. Sometimes I would look out at the birds I could see from this point in the moorland, allowing my attention to drift and wander. There were the swans and geese that swam upon Loch Dibadale; a squadron of terns that scolded and swirled above a tractor making its way along a stony peat road; lapwings that rose, fell and called out in shrill voices around the sheep fank, not far from the sign that marked the entrance to the village. A dark-winged bird would caw out and I would peer in its direction, wondering if it were a raven, rook or crow. Far from being a desolate, empty place, the moor seemed to have a thousand forms of life that called for my notice. There might also be the attentions of midges, the raw coldness of the wind, all biting me, gnawing through my bones. In short, it was not long before I would let the peat blade wander too, scoring my father's hand with its edge, drawing a thin line of blood.

'Keep it in,' Dad would call out to me each time his hand was cut. 'Take care. Keep it in.'

And then there would be the shortcomings in my wall-building abilities. I might manage to cut too many peats in a row. At other times, there would be too few. The 'blocks' of peat I made would be uneven; some would be too thick to make for easy drying, others too thin. Sometimes I would just slap one peat on top of another, making the entire structure liable to collapse at any moment. I would build it too close to the verge, sometimes too far away. My father's shake of his head would indicate how unsuitable he thought I was to perform any role in the construction industry.

'Obh ... obh,' he would utter a well-worn Gaelic expression of dismay.

He would stretch out an arm to keep the ridge upright, forcing each peat back into its proper place. Sometimes, too, he must have uttered a silent wish that none of the neighbours would see the sloppy work his son had performed out on the moor. Many of them were most particular about the neatness and precision of the task they had done. One woman was rumoured to go out to the moor with a cloth in her hand, cleaning away any bumps that had been left behind by the blade. In other houses, the lack of exactness shown in the cutting of peats seemed to be a topic of conversation for much of the following seasons, as much of a constant in their home as heat from the Rayburn stove.

It was probably worse when we reached the next level – the *caoran* or the *mòine dhubh* (black peat), known as 'blue peat' in Shetland, perhaps because of the oily sheen that glowed upon its surface. This was, as the poet laureate of peat, the late Seamus Heaney, once wrote, 'the good turf', the one closest to coal both in its shade and its heat-giving properties. We dug down to obtain this, sometimes four or even five layers, seeking the peat that was closest to the fire at the earth's core. Sometimes there were the roots of the old trees that had once covered the island found within the peat there, burnished silver by its oil. I would take a moment to examine it, glad of a moment's rest, before tossing it to the side. Despite these moments, there was little doubt that, working on a softer, easier substance, and like any worker eager to go home at the end of the day, people would work more quickly on this level than those who dug the more fibrous peat near the surface in the early morning. A Gaelic expression summed this up:

> *Chuireadh fear a' chaorain smùid à fear à bhàrr-fhaid!*
>
> The man on the black peat will always conquer the one on the first!

There were times, too, when it was not just one pair of family members who travelled out to cut peat together. A few or many might appear, performing what appeared to be a ritualistic dance on each layer of the peat bank. The first couple would step out on the top bank; the second follow in their paces; the third begin their task a short time later. And so the day would go round like an endless episode of *Strictly Come Dancing*, with each pair afraid to slow down in case they lost either points or position to those who followed. Their only reward for their performance was the tea, scones and cakes served out by the women of the household. Some would carry this to a ridiculous level, the cups and plates perched on a white tablecloth for grimy-fingered men to reach out and bring large slices of Victoria sponge to their lips. The liquid they served out from vacuum flasks and giant teapots possessed either of two shades. Sometimes it resembled *mòine bhàn* or the white peat they had cut earlier, possessing all the power and kick of the weak-kneed sparrows that used to gather round to peck at the available crumbs. At other times, it resembled the baths that greeted us when we arrived home, a bottomless well of diesel. This was especially true on days and nights when there was heavy rain, the filters employed by the water board unable to prevent large dollops of peat spluttering through taps.* As this tea was not for the faint-hearted, it was, perhaps, understandable that some chose to bring more reliable brews. Cans of MacEwan's Export – or 'Shetland Roses' as they are sometimes known in islands to the north, gaining that title for the way they bloom along the roadside, cast out empty from cars – were drained. The occasional whisky bottle would be brought along for the occasion. I have a childhood memory of my brother and I finding a bottle of yellow liquid somewhere near the boundary between our peats and a neighbour. Opening the bottle, we took a deep breath, inhaling the foul substance contained within. Our noses wrinkling in disgust, we decided to pour it away. It was a response that caused those working on the neighbouring peat bank consternation – and my father a great deal of money when he had to replace its contents at a time when whisky was dear.

* Whisky aficionados might be aware that – to the best of my knowledge – Highland Park is the only distillery in Scotland that still cuts its own peats, doing this in small quantities on Hobbister Moor between Kirkwall and the village of Orphir. Most others obtain the fuel for their process from the Northern Peat and Moss Company based near Peterhead.

After the cutting, there was a great deal of slow, tedious work. I recall spending hours on the moorland, turning the peats over to ensure they were dry on the underside. Later, we stood them up in *rùdhain*, small structures that resembled the mathematical symbol for *pi* with a number of extra legs attached, allowing the wind to rush through. Occasionally we might turn these small structures round, slipping off the 'hats' and replacing the legs in the opposite direction. Following that, there was the gathering: small mounds of peat beginning to appear on the edges of banks, slowly and painfully constructed. Sometimes this was done by wheelbarrow; on other occasions they were balanced on hand and arm and brought to the tiny stacks that rose there, waiting for the day the peats would be brought home.

As I look back, there is little about this time for which I feel any great longing. We performed this work under the combined battering of wind, rain and midges. As anyone who has visited the West Highlands, especially in July and August, knows, midges – members of the *Chiromidae* family – dance and swarm around the dampness of the moor. There were evenings when their biting, accompanied by the attentions of the even more unpleasant Messrs Tick and Clegg,* used to drive the sheep grazing in the distance back towards the village, bleating and racing towards the homes of their masters as if they believed human beings must possess a cure. There was no hope of that; their masters themselves scratched, scraped and performed their sole dance of the season as they attempted to be rid of them. For the greatest percentage of the year, they used to mock the gyrations of the dancers they viewed on our favourite programme, *Top of the Pops*. Now they sought to imitate them, stamping their feet like Noddy Holder,

* Not the politician and former Leader of the Liberal Democrats of that name, it should be noted, but an even greater irritant.

shaking their few grey locks like Marc Bolan of T. Rex. Even on dry, windy days, there was dust, whipped up from the surface of the moor. On certain days, our world resembled a darker, tan version of David Lean's film *Lawrence of Arabia*, with Peter O'Toole emerging from a desert of peat laid waste by a hurricane. I have written about this in a poem, referring to a story from another kind of desert land.

BY CANAAN'S SIDE

When we reached home from working in the peats,
it was like Jacob's household coming back to Canaan.
We removed all outer raiment, boots and socks from feet,
deplored the dark-rimmed nails upon our toes and hands,
and decided we could not begin to eat
till we had cleansed each molecule of moorland
from hair and skin. Only then we could take a seat
at table, join once again the ring of men
which we had left to go and seek
fuel for the stove that provided water, heat
to steam, hiss and bubble, clearing off the reek
of moor and heather, a wasteland fit only for deer and sheep
which we looked on as Jacob viewed the desert
from his fine croft in Canaan – wild, desolate and haunting,
threatening, bleak …

And then there was the peat fire flame itself, the generator – if books or CDs of accordion music were anything to go by – of a thousand tales and tunes. There are some who feel nostalgic for its scent, an understandable response when there is a slight whiff of its aroma in a room. It creates a cosy atmosphere, a warmth that curls and coils its way around you, as well as evoking childhood memories for me and my like. However, there are times when peat smoke can be both unpleasant and breathtaking. In the blackhouse – a thatched cottage with an attached byre, the earlier housing of the Hebrides – when it rose from small gaps in the thatch, there were times when it fouled the inside of people's homes and scarred and injured their lungs, creating major health problems.* The housing that the business magnate Lord Leverhulme saw during his time in Lewis and Harris clearly had a profound effect on him. His famous remark that their homes were 'not fit for kaffirs' was a reminder of how he had gained his wealth in the horrors of the Belgian Congo, as well as his shock that there was so much poverty still within the borders of this country at that time.

At its most extreme, the perils of peat smoke are generated by moorland fires, set off either by accident or – on rarer occasions – through the work of arsonists and others seeking to 'improve' the land for their own profit. The consequence of this could be seen in the summer of 2010 when dark and billowing clouds left the citizens of Moscow gasping for air, creating fatalities in Russia's capital.† Or in those nations suffering from the effects of Indonesia's peat fires in 2015, releasing more 'greenhouse gases' than all the cars and industry of Germany that year. I felt an immediate sympathy for them, especially when – for many of us – the thought of its choking power is not hard to imagine. It is, on a less dramatic level, a reek with which those of us of a certain vintage are all too familiar. Step into a blackhouse and we are reminded of its potency and power. It summons up an

* On the credit side, the layer of carbon dioxide created by smoke stopped the thatch catching fire and killed off much of the insect life inside the house.

† These events are not uncommon in places where there is a broad expanse of peat, such as Siberia and Canada. They have even occurred in California, where a peat fire in the Baldwin Hills area near Los Angeles burned underground for several years in the twenties. In tropical peatlands, the fires are often created by those who want to plant crops – such as oil palms – on a particular stretch of ground.

age when, if there was a quick change in wind direction, a gust might send smoke swirling into the kitchen, leaving everyone within range of the Rayburn coughing and spluttering incoherently. Arms would be flapped and cloth caps shaken in a mad, frenzied dance that someone had obviously designed centuries before to clear that smell and smoulder from the room.

There have been occasions in my life when a peat blaze has done more than that. I recall one or two instances when muirburn, a practice designed to improve the quality of grazing by setting fire to the heather within a patch of land, flared out of the control of the crofter or farmer who sparked it off, fire bursting out of its planned confines to threaten someone's home or croftland with destruction. This could sometimes happen through human error. There was one time, for example, when an enthusiastic individual managed to knock over a portable gas stove from the lochside where it had been placed. A few hours later and the fire brigade were hard at work trying desperately to dampen down the flames his small accident had sparked off on that dry summer's day. Despite their best efforts, it continued to blaze in the direction of the borders of the burgh of Stornoway. Smoke hung over the town with all the persistence of one of its customary rainclouds for days after that patch of heather was set alight.

My father would have found it disconcerting and disorientating for that smell to be there. When he was young, he apparently used to close his eyes every time he travelled into Stornoway. Breathing slowly and deeply, he reckoned he knew exactly the moment the vehicle crossed into town – when the aroma of peat was replaced by one that was alien to him, the smell of coal. With its presence, there was a sense not only of the town's boundary being crossed, but also of his entering into another form of experience, an urban world where people's lives were not limited in the way they so often were in croftland. The boy would have been acutely conscious of the differences between him and those who brushed past him in the street. They were familiar with trees and pavements, lampposts and large shops. They even spoke a different tongue from the one found upon his lips.

Yet even those of us who have reservations about peat were aware there was one day within the peat season when, if the weather was kind to us, our section of the moor could be described as festive. Tractors and trailers – sometimes a fleet of two or three – would rattle along the peat track towards the place where a large gathering of villagers had come together, waiting to fill up a load of the fuel they had cultivated so carefully for months beforehand. It was there that both I and others once again showed our more energetic qualities, throwing both *mòine dhubh* and *bhàn* up into the trailer to form a pyramid of peat, the load rising slowly from the depths of the trailer. There were times when we missed our target; Hebridean crofters were clearly not blessed with the accuracy of bowlers in cricket matches far to the south. Sometimes the heads of others might be hit. Sometimes, too, their hands would be scuffed and bruised. This was particularly the case when some of the younger and more eager scrambled up to stand at the crest of the load, making sure it was flat until the moment that it needed to be rounded off. This was to ensure that the trailer took a full load home, where the older people were waiting, ready to build a stack of fuel for the coming months. Sometimes, too, we were forced to empty it again. This happened when the great whirling tyres of the tractor broke through the turf, making it impossible to move forward with the load. We would gather together – young and old – shoving the wheelguards with our shoulders, trying to make it shift. Sometimes another tractor – or even more – would turn up to organise a rescue, pulling it out of the ditch created through the rolling of the wheel. There is a wonderful poem by the Shetland writer T.A. Robertson, known as Vagaland, in which he sums up in a chorus the mixed feelings a day of labour in the peats inspired:

> Oh, hit's truly odious lightsome
> When you're wirkin ida hill.

It was also the time of year when I gathered stories. I would listen to the older men talk about their travels, their days in the Merchant Navy, perhaps, or how they had been involved in the creation of hydroelectric dams in other moorlands on the mainland of Scotland. They would mention exotic locations like Cannich and Cruachan and how they had worked alongside 'Poles' and Irishmen there. Some of these migrant workers had volunteered to be Tunnel Tigers, taking explosives into clefts of rock in order to clear a passageway for other men and machines. Sometimes their bones and blood were still to be found there, tinting the stone on occasions when either the fuse or the TNT went wrong. They would shake their heads with amazement at the thought of them, especially the 'Poles' – a group that included Lithuanians and Ukrainians among others – who at that time had no croft or country to which they could return. 'They were desperate people,' one would declare. 'They needed the money to make new lives for themselves.'

There were other ways I gained enlightenment there. I remember being perched on a pinnacle of peats in the early seventies beside my fellow villagers, the Campbell brothers, and one of them – Alasdair, I think it was – pointing out a dark bird with white stripes on its underwings, dipping and scooping across the width of the moor. A fast, athletic flier, it was one I had never noticed before.

'It's an Arctic skua,' he declared.

A few moments later and I was treated to a lecture about the bird, which had only recently – like its relative, the great skua – come to nest near the top of Lewis, being absent from the Hebrides before. The two men spoke about some of the more unpleasant habits of that creature, how it swirled around the heads of oilmen in the North Sea, scraping their helmets with its beak, forced seagulls, terns and other birds to spit out the food they had consumed a short time before, swallowing it whole. After that, the conversation shifted to more weighty concerns, such as how a ship gained entrance to Sydney or some other harbour on the far side of the world. Some of the men could become strangely vociferous discussing the navigating of such channels, convinced that they – and only they – were right.

It was not the only quarrel that had taken place in the moor near our home. Once I recall working with my father on the peat bank when an angry neighbour strode towards us, accusing us of cutting the peats on the edge of his bank, approaching the boundaries of his domain with *tairsgeir* in hand. My father listened while the old man fulminated, never pointing out that it was also the case that his peat blade was moving in the direction of our fuel supply, staying silent, too, about the likelihood that – with their respective ages – the two of them would no longer be in need of any peats by the time the two banks came close to touching. Eventually, the other man stamped away, feeling certain that his point had been made, fuelled by a righteous anger.

Such 'turf wars' were not rarities, that upright fury almost the standard way of behaving when peat was involved. There is a story in the novel *The Fish Can Sing* by the Icelandic Nobel prize-winning writer Halldor Laxness which shows a man walking to his neighbour Bjorn's home in Brekkukort, the novel's central location outside Reykjavik, with a sack over his shoulder. It contains peat that he has previously stolen from Bjorn and is now, with a bad conscience, attempting to return. The crime is a grave one, especially in an area where fuel is carefully husbanded. A shortage can either kill someone with cold or leave him without the means to cook food. His original act is seen as even more heinous because Bjorn, a character known for his compassion, had earlier given him peat. For an Icelandic reader, Bjorn's generosity and forgiveness were the ultimate acts of mercy.

It could be argued that it was out on the moorland where the average Lewisman indulged most often in at least two of the deadly sins – wrath and greed. Sometimes this was about the direction in which sheep wandered and grazed. Some crofters appeared to believe that certain areas of moorland were theirs by long-held and mystical rights, bestowed by the manner in which their flocks kept returning to particular areas of the moor year upon year. In English, this is termed the 'hefting' of their sheep, and was used as a justification to enable some of these individuals to find both ways and means of preventing other people's flocks from entering their 'patch'. There were times when this was expressed in the need for an apportionment. Generally, this found its expression in an area of rough grazing on the edge of the moor, one with little value apart from the fence wire that surrounded it. Its existence also left a legacy of bitterness in the community. Brothers would not talk to brothers on account of an acre or two of ground that was colonised by rush and bog within a year or two. As a result, there is much with which I can identify in the words of the Irish writer Patrick Kavanagh's short poem 'Epic', when he writes of how:

> I have lived in important places, times
> When great events were decided, who owned
> That half a rood of rock, a no man's land
> Surrounded by our pitchfork-armed claims.

Most often, however, it was the peat bank that was the centre of dispute. One of the *Stornoway Gazette*'s more interesting headlines, way back in 1955, reads 'Brother and Sister Quarrel about Peat Banks', detailing the ferocious arguments that led the two siblings to an appearance in the Land Court. There is a story from 1963 in which a

51-year-old unemployed man from another village is found guilty of 'wilfully and maliciously damaging a quantity of newly cut peat', a most heinous crime in the eyes of the community. People probably didn't speak to him for years afterwards. Apparently, he was sentenced by the judge to go out and cut new peat for his victim.

One observer also noted with concern the time a few years ago when the Irish government introduced a measure to tackle that nation's economic woes. This was a £15 per tonne carbon tax on various fuels, such as domestic oil and coal. More worryingly, it was even being applied to the ordinary household peat that kept the home fires burning. Certain individuals in the islands in the north of Scotland suffered nightmares at the thought of the UK government ever introducing such taxation on one of the staples of life. They had visions of HMPs – or Her Majesty's Peat-inspectors – standing at the end of every village road, checking both the quality and quantity of the fuel brought in from the moor on trailer and tractor. There might even have been two rates at which the tax could have been levied. The special dark (or blue) peat – or *mòine dhubh* – would have been subject to a higher rate of payment; the lighter, whiter and more fibrous material valued at a lower percentage of its price.

In line with this idea, Her Majesty's Government could also have brought in a PAYB – or Pay As You Burn – scheme. This might mean specially disguised members of the Peat Fraud Squad visiting homes across the north of Scotland. They would arrive in the thick coats and bonnets of a spinster aunt or, perhaps, the cloth cap and tweed jacket of an elderly uncle. It would not take too long for one to be hoodwinked by their good cheer and company, perhaps eased into false comfort by the whisky or sherry they had brought along for their visit. The householder would not have realised that each time they are stretching out to feed another peat into the flames, their guests are undertaking a little mathematical challenge, trying to measure the size and scale of the peat you are placing in your Rayburn. It is only when the visitor heads out the door of their homes that the hosts realise the exact nature of the mental arithmetic the old man or woman has been performing, attempting to work out just how much peat has been chucked into the flames during the average week/month/year, depending on the warmth or otherwise of the season. One other notion would have been the introduction of VAT or Value Added Tairsgeir/Tushkar/Tuskar. This would have been a source of revenue determined by the size both of the blade and the peat it produced. As such, as critics have pointed out, it would be a highly discriminatory tax. Shetland's tushkar, for instance, has a small, feathered cutting edge at its base, the Lewis variety a much larger one.

During my teenage years, however, the idea of a Peat Inheritance Tax would have concerned my fellow-villagers most of all. The size of the stacks beside the houses sometimes illustrated a certain tendency among my fellow Hebrideans. Despite the fact that they had more than enough fuel beside their homes to see them out for several lifetimes, some people still seemed to want more, one that might warm and comfort them during the next Ice Age if any happened to be impending. It was a trait that was most pronounced in my home district. The island humorist and a gentleman from the more southerly parish of North Lochs, Hector Macdonald, in his weekly column for the *West Highland Free Press* once announced his ambition to picket all the houses in the Ness district. The purpose behind his action was to act as the local representative for the Society for the Prevention of Arson, protesting against the size and scale of the peat stacks in the area. He envisaged this group picketing the parish's households, waving placards that probably bore slogans like *Have a Hearth – Stop Burning Peat! Douse Your Flame Now – Or One Day You'll Come To Feed Them*, or even *Down With This Sort of Thing*. For all his humour, there is little doubt that the man had a point. Several of those with huge stores of *mòine dhubh* and *bhàn* beside their home had a fondness for the peat-fire flame that bordered on clinical obsession.

Yet there are occasions in his book *A View from North Lochs,* based on columns he wrote in the early eighties, when Hector betrays a similar passion for peat. He describes a series of nightmares he experienced when in the hours of darkness he finds himself 'in the black depths of a three-deep peatbank', accompanied by no less a figure than Margaret Thatcher. She apparently had come 'walking across the Minch, golden mane flying in the wind, carrying the *tairsgeir* in one hand and a whip in the other'.

Allegedly she sings, too, various renditions of 'Onward Christian Soldiers' and 'When The Saints Go Marching In', urging him onwards. Perhaps a more fitting suggestion would be the words of a Gaelic song we bellowed out in the Ness Hall of our youth. In shades of Edward Lear, one can just imagine the words endlessly repeated on the late Prime Minister's lips:

On a man who is late in marrying	Fear a bhios fada gun phòsadh,
Grass and heather and moorland plants will grow;	Fàsaidh feur is fraoch is fireach air;
On a man who is late in marrying	Fear a bhios fada gun phòsadh,
A great beard will grow.	Fàsaidh feusag mhór air.
On a man who is late in marrying	Fear a bhios fada gun phòsadh,
Grass and heather and moorland plants will grow;	Fàsaidh feur is fraoch is fireach air;
On a man who is late in marrying	Fear a bhios fada gun phòsadh,
A great beard will grow.	Fàsaidh feusag mhór air.

This strange obsession seems to have been even more intense further north. In *Humours of a Peat Commissioner,* we have no fewer than three volumes written on the subject by Thomas Manson, a Shetland writer who published the first section in 1918. Much of the work is dialogue, often exchanges between the Literary Member and the Practical Member of the Commission, a body set up 'ta fin oot aal aboot paets in Shetlan', including where they are, where they are not, the price they are, how long it will last, the size of it... The Literary Member is, as befits his name, concerned with the poetry and magic of the peat, regularly concerning himself with such important matters as whether or not Robinson Crusoe harvested peat while marooned on his island. The Practical Member, Jeremiah Laurenson, has much more mundane matters on his mind. Responding to a song that contains the lines:

> In bleak, cold December, it hails and snaws,
> If the fire goes out, his fingers he blaws

he comes to the following conclusion about the singer: 'He's shurly an Engleesh body, whaar dey use col. If he'd come to Shetland, he widnae hae cauld fingers fur he wid hae plenty of paets.'

Despite my reservations, there are reasons for this odd behaviour, which now as an adult I can understand. Brought up in the old blackhouse without electricity until the fifties and, in certain circumstances, a decade later, many of my neighbours and the majority of people in places like Shetland had depended on peat for much of their lives. It was the provider not only of heat but also the means for cooking food, either above an open fire or the kitchen stove. In certain circumstances, with the narrow windows of their houses, it was even an important source of light. These people knew want and hunger, suffering this as late as the early twenties, when famine relief – in the form of venison – was delivered to the people of my native district of Ness. They would also gain the assurance that if the following year was damp and wet and they were unable to harvest enough fuel, they still possessed enough to keep them warm and dry for a time. As such, peat was a precious commodity, one which had not only to be hoarded but also treasured, treated with a degree of delicacy and care.

There was even a little evidence of that in the peat stack itself, especially in the meticulous way in which it was constructed. Neat, ordered and harmonious, it has become recognised over the last few years to be an astonishing example of folk art, exhibited in An Lanntair, the Stornoway Arts Centre. With their geometric patterns, the stacks formed much of the traditional architecture of the Hebridean village, an expression of man's need for both precision and beauty. Sometimes the outside of the peat stack formed straight, symmetric lines. At other times, it resembled the herringbone tweeds that some of the island weavers had stretched out on their looms. They offered a similar warmth when they were wrapped around human flesh in the chill of winter, when people made small, continual journeys between their supplies and their home-fires in order to gain warmth and comfort. Yet small disorders and avalanches were created each time a peat was tugged out, the tidiness of the stack crumbling in a slow and continual decay.

CHAPTER FIVE

Grauveen (Dutch) – Grey Turf

There is much of the Ness moor that resembles the setting of a John Ford western. Through half-closed eyes, one can see the landscape of such films as *Fort Apache* and *She Wore a Yellow Ribbon* in all their emptiness and desolation. Even the shade of John Wayne – especially with his real surname, Morrison, just about the most common in the district – could be summoned in the shape of some of the local crofters looking for a missing sheep, taking on that role instead of playing Ethan Edwards in his search for his niece Debbie, held captive by the Comanche. This similarity was most pronounced in a place called Gleann Dibadale, a heather-topped valley at the edge of the village. Here were hills that young boys could race down, hollering war-whoops as they went. Here were jutting rocks and precipices, which we could lie behind and try to bushwhack our friends. Here, too, was a substitute for the OK Corral with walls made of turf* built to gather sheep from the moor. We would perch behind it bearing imaginary rifles, draw out pistols constructed from sheep horns, making loud gunfire noises with our mouths.

'*Bwaamm … Bwaamm … Bwaaammm …*'

One of the aspects that lent some reality to that scene was the fact that, many hundreds of years before, there had been a real-life skirmish in the valley, a battle between the Macaulays, who lived in Uig at the southwestern end of the island, and John Wayne's old clan at its northern tip. Somewhat in the spirit of the western, the clash had allegedly occurred as the result of some cattle-rustling in the area.

* Sometimes walls of turf like these formed the foundations of the older houses of our villages. A way of saving money when they were first built, they were later said to cause dampness in a home.

The graves are still there to this day, or so we imagined as village boys, heading out there with our fathers' spades one afternoon, leaping up and down on their hilts, turning over turves to see what lay below. Our expedition was largely the result of seeing small, man-sized quilts of heather at the edge of Loch Dibadale, a stretch of water that lies at the more fertile, greener end of the glen, used as richer grazing for the cattle belonging to the people of South Dell. Urging us in its direction were rumours of treasure. This came in the form of a clutch of discoveries there during the course of the twentieth century. There was some jewellery: three silver arm rings and two silver finger rings from the tenth century, found by one of the women from the village in 1938. Nearby, two Bronze Age swords, from the late eighth century BC, were also discovered within a few yards of each other, their hilts jutting out from the edge of a peat bank. One of these was in good condition – sharp, complete and with a bone grip. I have seen it in the National Museum in Edinburgh, its shade a rich brown, coloured by the peat in which it had been buried.

Yet in some ways, it wasn't the swords that brought us there, scrambling hopelessly with our spades, but the legend of the man who discovered it, one who had noticed its gleam while, allegedly, standing by the peat bank relieving himself. Murdo Maciver was – even by the standards of a rural life at that time – one of life's eccentrics. Though he had died some eight years before I was born, there was still a legion of stories in existence about him and his brother, Donald, who lived at almost the last house on the northern end of the village some time later. Both men were bachelors; both worked as joiners; both, too, probably suffered from shell shock as a result of being involved in the Boer War. The two also had another trait in common. Their path through life was continually diverted by the presence of fairies. They would chase them with fishing nets taut between their fingers. Murdo would walk from the local well with a pair of enamel pails in his fingers, before abandoning and upsetting the buckets after seeing a couple of sprites in the distance. At other times, he would race through a field of oats with a scythe in his hands, trying to chop the slight and elusive figures down.

Focusing on this, however, is to deny the man's complexity. Murchadh a' Chlaidheimh (Murdo of the Sword, as he was known by his fellow villagers) was a man of quick and sharp wit, his eye often sharp on the foibles of his neighbours. He wrote many songs about incidents in the parish, most notably one shortly before his death in the late forties. This was the time when mail-order firms began to send out their wares in the post to the island. He scribbled the song 'Oran JD agus Dallas', effectively a catalogue of the changes of behaviour that the presence of mail-order firms like JD Williams had on the minds of the women nearby. He also encountered a man they nicknamed 'Tiger' in the blacksmith's workshop in the village of Habost a few miles away.

'*Agus an de' seo Murchadh a' Chlaidheimh?*' the 'Tiger' declared. 'And is this Murdo of the Sword?'

'*Nach fheum duine chaidheamh nuair a tha e coinneachadh tìgear?*' Murdo responded. 'Doesn't a fellow need a sword when he meets a tiger?'

Stories like these still glittered even decades later, bringing small boys with spades in their grip to the moorland to search for weapons, inspiring long conversations about buying metal detectors which we – with our pocket money – could not possibly afford, ignoring how we lacked the patience and determination to wield them even if they ever came to hand. Instead, we were forced to be content with the tangled roots of silver birch we encountered when setting out to dig peat a handful of years later. There were others, however, who discovered evidence of the island's past life when they were out digging peats in the Ness moor. An unusual, white flint arrowhead, unlike any others found in the island, was discovered near Dell in 1985; a wooden bowl from the Iron Age on another occasion. A horde of bronze tools, as well as some beads, fashioned from amber, glass and gold and dating from the seventh century BC, were found near the village of Adabrock by my namesake in May 1910. I recall, too, a body found by peat cutters in Habost 78 years later in May 1988. Its blackened bones were twisted and deformed and belong to a probable male of 15 to 18 years of age who was approximately 5 feet 5 inches (1.65 metres) tall.

In Airnish, another part of Lewis, in 1964, the body of another young man was uncovered in the same way. Like much that is discovered on the moorland, it is certain he was not intended to be found. On this occasion, however, the corpse – for all that its bones were reduced to the consistency of rubbery seaweed by this time – appeared to be the victim of a crime in the early eighteenth century. The back of the skull had been damaged. According to Dr Fiddes of Edinburgh University, 'the position of this fracture would be consistent with a blow with a weapon wielded by a right-handed assailant attacking his victim from the rear'. Rev. William Matheson of the same university's Celtic department tried to play detective in the *Stornoway Gazette* of July 1964. Noting that within the young man's belongings was a quill used for writing, he referred to an old Lewis story whereby:

> 'Two youths attending the school at Stornoway went to the moors on a birds' nesting expedition. They quarrelled when sharing out the spoil, and one of them felled the other by a blow on the head with a stone. When he realised that his companion was dead he buried him, and fled to Tarbert, Harris, whence he made his way to the south and took up a seafaring life.'

Despite the long, steep walk south to Harris, the murderer did not escape. Time and temptation took him many years later to Stornoway, where his ship tied up at harbour. He decided to go ashore, probably believing his identity had long been forgotten. But he was wrong. Some local recognised the stranger, a look or a gesture giving him away. Shortly afterwards, he was convicted of murder, and hanged on Gallows Hill, which was now within the castle grounds outside the town. It has been suggested in one version of the tale that he was buried beside an unusually shaped rock, known as Creag a' Bhodaich, which is not far from where the body was found.*

And so it goes on until it seems that no deed, no unexpected demise is safe from the attentions of the peat cutter. They were at work, too, in the northwest of the Shetland mainland, leaving no turf unturned as they uncovered someone who has become known as Gunnister Man, taking his name from a nearby deserted village. Dr Carol Christiansen, the Textile Curator from Shetland Museum and Archives, spoke to me when I visited her in the museum store one day. From Seattle in the United States, she arrived in Shetland by an unusual route, having studied in Manchester for her PhD, specialising in archaeological textiles. Her soft American accent stayed with her, like lightly carried baggage, despite all her years away. It is her task to introduce this individual to me, one that was removed from what turns out to be his second grave in 1951. His corpse had been found before, possibly a short time after his death, yet already 'too decomposed to be taken to Ollaberry or Hillswick', as Carol pointed out were the nearest graveyards. Instead, he was reburied a short distance from where he died, the bucket he carried left below his feet, his walking stick placed across his chest. How or when he stumbled is one of Shetland's greatest puzzles. There were no tales about a dead man lying in the vicinity

* Perhaps the most beautiful object I have seen was a wooden toolbox found at Birsay at Evie in Orkney. It contained a number of leatherworker's tools, and probably belonged to a professional leatherworker sometime between 640 and 860 AD. It was discovered in the nineteenth century. Rather unexpectedly, the box has carved decoration on the outer faces, a rare surviving example from Scotland at this period.

before he was found. Some of the local children even used to stand when they were waiting for the school bus in the morning on the mound where he was discovered.

'They wouldn't have done that if there had been any stories,' Carol said, smiling.

Much that is mysterious still clings to him. He was found in a time before DNA had been discovered, making it impossible to be certain of the date he hiked across to Nibbon, the nearest community to Gunnister. Since that time, too, his bones have been lost somewhere in the National Museum of Scotland. Even before that, only a few fragments of his body remained – bone, dark hair, fingernails and toenails. If this was all that had been discovered, he would have been long forgotten.

In this case, however, as Dr Christiansen explained to me, 'clothes maketh the man', his outfit telling us something of his personality and, perhaps, his past travels. In some ways he seemed to be, in the words of Ray Davies of the Kinks, 'a dedicated follower of fashion', one who imitated the styles and modes of dress of the late seventeenth century that swirled all around them in that northern edge of the world. His long coat and trousers possibly came from the Low Countries, resembling those worn in Marken in North Holland, which was, until 1916, an island within the Zuider Zee. (The coat even had false pockets, allowing damp from rain and snow to soak his upper legs.) On his head were two caps, one without a brim and knitted in an elaborate pattern, most likely from Norway or Sweden; the other simpler and more straightforward. His purse, which contained both Dutch and Swedish coins, had a pattern similar to the Fair Isle knitting that is still found in the islands today, consisting of two colours. His gloves, too, had little decorative shapes stitched on their upper palms and elsewhere. Nearby was a leather belt, a silk ribbon, a horn spoon and knife handle.

There is little that can really identify this individual, no clue that might inform us even if he was a native Shetlander or a visiting trader, perhaps, from somewhere in the north. Even his age is something of which we are unsure. Yet there is enough here to build some sense of this individual as an identifiable human being. Visit any town or city in our country on a Friday or Saturday night, especially in wintertime, and one can see the modern-day equivalents (both genders) of the Gunnister Man. They step to the left, step to the right, wearing garments that are too flimsy for the chill of wind or sleet, believing that style and not warmth is of ultimate importance. It is, perhaps, a comfort to us that this is not just a modern phenomenon. Someone like the Gunnister Man behaved like this in the mini Ice Age of the late seventeenth century, years when the River Thames lay frozen and when death by hypothermia was all too familiar. They strode out in clothes that were thin and inadequate despite the risk of blizzard or snowstorm. It might even have been the cold that disorientated him, making him step away from a path that might already have been blurred by a flurry of sleet, a thin covering of snow. It may have been tiredness. In the end, it does not matter. He became yet another casualty of the times, one in which, according to Elizabeth Ewan and Janay Nugent, 'Famines in France 1693–94, Norway 1695–96 and Sweden 1696–97 claimed roughly 10 per cent of the population of each country. In Estonia and Finland in 1696–97, losses have been estimated at a fifth and a third of the national populations, respectively.' Shetland, as the Gunnister Man's clothing might suggest, was a close neighbour to a few of these countries, sharing a similar toll of deaths.

Yet there is much more than fashion that peat preserves. Sometimes, as Dr Christiansen showed me, it can be a tiny whorl of wool, dropped into the peat, perhaps, by an individual working there some centuries before. It can be bog butter, its name belying the fact that it is not always a dairy product but sometimes adipose or sheep's tallow, a waxy substance conserved by being placed in a bog to prevent its rotting. The cool temperatures and highly acidic nature of the land down below the moor's surface may have done this, a similar process to the more modern art of the fridge-freezer.

For this reason, items like butter were often preserved in wooden containers or even within the skin of an animal like a deer. This had the additional advantage of allowing food to be both hidden and stored, especially if thieves, berserkers or raiders happened to be roaming around. This is a practice carried out since the second or third century in Ireland and even before that in other countries like Scotland. It may even have been a local speciality, with butter, for instance, being drawn out of its place within the moorland at certain times of year. The English writer Thomas Dinely in an account of Irish life dated 1681 describes a special feast that was served for him. It contained: 'Butter, layed up in wicker baskets, mixed with a sort of garlic and buried for some time in a bog to make a provision of a high taste for Lent.'

Others have suggested that it is the transformative qualities of the peat bog that were most useful. Burial within the moorland was a step in the preparation of some highly perishable foods, such as meat and dairy products. A form of cold storage, it may have done more than delay the rotting process occurring in certain foodstuffs. On Islay in the Inner Hebrides of Scotland, for instance, where seabirds were often the daily dish, any gannet caught and designed to be dined upon was first buried in a peat bank for ten to 14 days to remove its gamey taste. This technique was especially important in some traditional cuisines, particularly in areas at some distance from the shoreline that lacked a ready supply of salt. This was true in the north, as in the case of *gravadlax*, the Scandinavian dish cured in salt, sugar and dill, or *hákarl*, treated shark from Iceland, and the Inuit dishes *igunaq* (walrus and other sea mammals fermented and frozen in snow) and *kiviaq* (auks left to ferment for months inside a sealskin). My own ancestors probably used culinary techniques like these. They certainly buried fish such as skate, often in the household manure heap. It was, I have to say, a taste that passed both me and my entire generation by.

And these discoveries still continue in peat bogs, especially in Ireland. Late one morning I scrambled down the banks that led to a stretch of waters known both as Poulaphouca Reservoir or Blessington Lakes, not far from the township of Blessington in County Wicklow, one of the peat counties in the centre of Ireland. I was accompanied by two men, the musician Kevin Conneff and his friend CJ Darby. We made an unlikely trio – me lunging from time to time, trying to maintain my 6 feet (or so) upright on a grassy slope, imagining the cries of 'timber' if I was ever felled; CJ probing out his path with the tip of his walking stick, looking carefully at each step; Kevin a jaunty-looking figure with red jacket, blue jeans and a dark blue Dublin GAA cap. We were going to places where, judging by the beer cans and empty WKD bottles, few grown-ups had gone before. A tangle of stalks and branches and the occasional skid of mud seemed set in place to hinder us on our way, obstacles deliberately fixed for those who were no longer adolescents.

Created by the construction of a dam in the late thirties, the depths of the reservoir concealed – among other structures – the townland of Ballinahown, its lights now permanently dimmed by the coming of that rare phenomenon in Ireland, hydroelectricity, to these edges of the country. Most of the time, the name Poulaphouca, the Irish translation of the term 'Demon's Hole', seemed a perfect fit for these waters, which had swallowed all memory of the existence of the settlements. Occasionally, however, as the water ebbed and flowed, there would be small glimpses of the past. A road leading to darkness. Four concrete pillars on which a haystack once stood. And then, too, there was evidence of even older life than that. On the edge of the shore were what appeared to be dark bruises in the gravel of the shoreline, small residues of the peat that had once lain there.

'Do you see anything there?'

CJ now showed the reason for both his slow, careful pace and the walking stick he clutched in his hand. He was pointing at a dark area at the base of a small bush not far from where the waters lapped.

'Well?'

For once, I was the one who saw it first – a stretch of white that was neither stone nor root. A moment later, Kevin spotted a similar pattern elsewhere, slightly further back in the shadows.

'What is it?' I asked.

'It's the bones of a large deer embedded in the peat. The one they used to think was the great Irish elk.'

'It isn't...?'

'No. Nowadays they think it's a relative of the red deer and not connected to the elk in any way.'

I felt a small disappointment, recalling the description of that animal in Heaney's poem 'Bogland' as a 'crate full of air'. I had seen, too, the skeleton of one perched in a museum in the German city of Bremen, defying gravity on its plinth. In comparison with the anorak made of fishskin and the flightless Kakapo parrot that were also within the building, it looked refreshingly normal.

'I found an antler there a little while ago. I brought that into the National Museum and let them know the rest is here. They'll be down here some time to have a look at it.'

Both Kevin and I grinned, delighted at his good friend's discovery. It seemed further proof of what Fiona Coates the Australian botanist had claimed, that wherever it existed, peat was like an encyclopaedia, one that contained much knowledge of the past. She had seen it in terms of the pollen found within the bogs. It was a notion that also applied to animal life and even to the existence of humans who had walked on the planet before us.

Over a decade ago, in 2001, while I lived and worked in Benbecula, one of the Western Isles, I saw proof of that. Together with my son and daughter, I walked out on the western edge of South Uist one Sunday, not far from the village of Daliburgh. I was aware there was a team of archaeologists from Sheffield working there, on an afternoon when wind and rain possessed as many spikes as the marram grass that protects the west coast of these islands from eroding and being blown away. They had scraped away sand to uncover a row of roundhouses, built – around 1000 BC – like a Victorian terraced street in, say, Lancashire, to share the same stone walls, each one gathering strength and shelter from the other. In one of these buildings there was a variety of strange-looking beings that would not have looked out of place in a cross between a horror show and an early episode of that long-running soap *Coronation Street*, all circling a table in the Rover's Return. One of the young archaeologists spoke to me about the find, shaking both because of the edge of the wind and her own excitement.

It turned out there were four complete skeletons at the site. One of them was a male whose life had come to an end *c.* 1600 BC. Another turned out to be a younger female who had died some 300 years later, around 1300 BC, buried along with a sheep. There was a 10–14-year-old – probably female – and a three-year-old. What was particularly odd about this was that the two adults had been fixed into a flexed position with legs bent and brought close to their chests. They had clearly shared their room and space with those who lived on the site, each day apparently a rerun of *Meet the Ancestors*, a TV documentary series that was popular in 2001.

Yet life – and that discovery – grew stranger still. It transpired that both these adult bodies had not been buried until 1120 BC some centuries after they died and, just as strange, they had been preserved in a quite different manner from those found in Tutankhamun's tomb, having been left in a peat bog for about six to 18 months. (Unlike their Egyptian equivalents, the soft tissue of the two bodies had not been conserved when they were found.) They had then been left in the roundhouse for reasons that are still unclear – though some have suggested religious significance. What made it odder still was that these male and female skeletons did not consist of two bodies. Instead they came from six three in each instance – with some of them from different eras. The jawbone of one originated from around 1440 to 1260 BC. Other parts of the head came from 1500 to 1400 BC. There was also a considerable quantity of smashed crockery and broken animal bones around the dead, suggesting some spiritual practice had been involved in this ritual.

There was a similar line-up of the bizarre and grotesque when I visited the National Museum of Archaeology in Dublin. In an exhibition entitled Kinship and Sacrifice, there were many items that had been scooped out of the bog that is 17 per cent of Ireland's surface.* They included not only gold jewellery, headdresses, horse harnesses and Viking swords, but also the discovery of a few of around a hundred bodies. Of all ages and both genders, some are like the Gunnister Man stumbling, perhaps, into a peat bog, or the man from Airnish on my island home, placed there in the hope that they will never be found. (This was particularly true of the recent dead buried in the Irish moors, those like the Disappeared, killed by terrorist organisations on both sides, Irish Nationalists/republicans and Unionists/loyalists, during the Troubles in Northern Ireland.) Others were given formal burials – for some reason preserved in the kind of earth that ensured they would be recognisable a couple of generations later.

And then there were the other disturbing ones, those like the corpse they call Oldcroghan Man, who lies under subdued and respectful lighting in a glass case in the museum. His stretched-out, bronzed body inspires more reverence

* This is the third-highest proportion in the world, after Canada and Finland.

than fear, apart, that is, from a small boy who stood near me with his father on what was evidently a male day out.

'Dad … it looks like toffee,' he declared.

His father hushed him, but not before the child's words slipped out from his mouth into the silence. We both exchanged quiet smiles as I thought of how right the boy was. There was something about the body that resembled the Barratt Everlasting Toffee Strip I used to grind with my teeth around a half-century or so ago. Over time, the toffee strip would disappear, being considerably less 'everlasting' than the body – some 6 feet 6 inches in length – that was stretched out before me. It was the arm and wrist of that headless, legless corpse which inspired the greatest fascination for me. His hand looked refined and delicate, clearly not one that belonged to these people whose role it was to dig or harvest swathes of County Offaly where he lived. Nor one that people anticipated from the remainder of his brutalised body. One could imagine his fingers signalling attention, snapping to bring some servant to his heels.

There was, however, more evidence about his life, one that is believed to have met its end after a stab to his chest, a blow that he may have tried to stop by shielding himself with his arm. A small wound there provided evidence of this. After being stabbed, he was decapitated and his body cut in half, severing his legs. This was not, however, the only evidence of violence. There were also deep cuts under each of his nipples. Several ideas have been put forward to provide an explanation for these wounds. Some have argued that the man was tortured while he was alive. Other claimed that these marks were caused, after his death, by conditions within the bog. More convincingly, there is the argument that the man was a king, sacrificed to the goddess of fertility for the failure of the harvests under his reign, which took place sometime between 362 and 175 BC. It has been suggested by Eamonn Kelly of the National Museum of Ireland that his killing was a deliberate act as a response to the coming of famine to the part of the country which he governed. Kelly described this sacrifice as a 'threefold death', one that often involved stabbing, drowning and the trauma caused by being struck by a blunt instrument. In addition to this, Oldcroghan Man also had a spancel fastened to his leg, the noosed rope that is used to hobble a horse or cow, and withies – the tough flexible branches of the willow tree – were fastened round his neck, choking him to death. It was an event that probably took place at one of the holy times of the ancient Irish world – Samhain or Hallowe'en – and deliberately within a stretch of bogland, as these were regarded as locations that, being half-water, half-earth, permitted entrance to the supernatural world. In contrast, areas of forest concealed the skies with leaves and branches, excluding entry to the upper world.

It was likely only after his execution that the man's nipples were defaced and disfigured for symbolic reasons. This showed that he was a king who had been rejected by his subjects, people who at one time had kissed that part of his body as an act of homage and obeisance – a ritual described by no less a personage than St Patrick, who had refused to kneel and perform this ceremony on a pagan ruler. That bond between governor and governed had now been broken and set aside, his nipples mutilated as a result, his broken body scattered.

It might suggest why he was found in the place he was as recently as 2003 – on the boundary between the two ancient kingdoms of Tuath Cruachain and Tuath na Cille, his dismembered corpse scattered to mark out the division between two different stretches of land, sacrificed, perhaps, to a goddess that represented not only land and fertility, but also sovereignty, war and death. Underlining the power of this figure is a range of objects found in places like these. They include bridle bits, wolfskin capes and shields, cauldrons, drinking vessels and even block wheels at a boundary at Doogary More in County Roscommon, all elements associated with a royal inauguration. Near where Oldcroghan Man lay in what is even today a border between two neighbouring parishes, there was a medieval castle belonging to the O'Connor family, the former rulers of Ui Faighle, a Gaelic-Irish kingdom that survived until 1550. Its name survives to the present in the name County Offaly.

Yet Oldcroghan Man is not alone, neither in the museum nor the fact that he was only recently discovered in the peat bog, not far from a boundary. Nearby, there is Clonycavan Man, also found in 2003, this time a short distance away from the border that lies between Meath and Westmeath, which separated at one time the kingdoms of Brega and Mide.* It is not far from a hill that was reputedly used for ceremonies involving kingship. Discovered in a peat-harvesting machine, which may have cut away the lower half of his body, this figure was found, too, to have been a victim of violence. A sharp implement of some kind, possibly an axe, caved open the top of his head, leaving his brain uncovered. There was also a large wound across the bridge of his nose, leading to his right eye – the blow that is believed to have killed him. Again, there seems to be a suggestion that this was a ritualised killing. His nipples were pinched and cut, rendering it impossible for them to be greeted in the proper regal manner. He had also clearly been wealthy. An examination of the tight black curls that were raised on the top of his head suggested that he had eaten a rich diet in the previous year. He also seems to have used an unlikely brand of hair gel, consisting of plant oil and pine resin, imported from either south-west France or northern Spain. It may even have been that, unlike his counterpart in Oldcroghan, he needed to appear taller than he was. Being

* Others, like those found in Derrymaquick, County Roscommon, Kinnakennelly, County Galway, and Baronstown West, County Kildare, were also found near the boundaries of old barony lands.

only 5 feet 2 inches, he might have required all the cosmetic assistance he could muster.

There are others that have been found in Ireland. They include two who accompanied the pair in the museum in Dublin, Gallagh Man from the Galway area, who was apparently strangled, and Baronstown West Man from County Kildare. There are also absent friends, not on display, such as Cashel Man, believed to be the oldest surviving body in the world today. Found in Cashel Bog in Ireland's dark centre in County Laois, he was believed to have walked on the Earth 4,000 years ago. Of a similar vintage might be the Moydrum Man. He was found in Rossun Bog in County Meath in 2013, dug out from the earliest levels of the bog. All of them illustrate what is the central paradox of the bog body. Placed within peat and marshland in order to be destroyed, it is that very element which ends up preserving them, even keeping and maintaining their very humanity. The victims, perhaps, of the cruelty of those who lived in the Moydrum or Oldcroghan area centuries before, their images and appearance have outlived those who ended their existence, haunting those who come to see them in locations like the National Museum of Ireland.

It is likely that some day soon the Moydrum Man may join the ranks of well-known bog bodies, from Lindow Man – or 'Pete Marsh', found in Lindow Bog in Cheshire – to the Grauballe Man and the remarkably preserved human face of the Tollund Man in Jutland in Denmark. It is clear that many such bodies are yet to be discovered, for as Heaney would have declared, 'the wet centre is bottomless'. Not in the literal sense as the Irish might once have maintained, but in the sense of its mysteries being unresolved and unfathomable. There is no end to the secrets peatland contains.

CHAPTER SIX

Skyumpik (Shetland) – Mossy Peat

It is hard to keep up with a man who plays the bodhrán.

That thought kept running through my head as I followed Kevin Conneff up the slope in the townland of Upper Lugglass near Hollywood in County Wicklow. His legs were moving as quickly as his fingers when they clutched the *cipín* (tipper), tapping out a tune for the Chieftains, the band in which he has played for the last 40 years, accompanying them all over the world. For all that he was ten years older than me, he possessed much more energy. He clambered over gates, lifting up the two small black-and-white dogs that walked with us as he did so, while I panted. They raced and rushed as we continued to climb the hill, shadowed by the dense plantations of Sitka spruce that now dominate the forests that make up around 10 per cent of the Irish countryside. The government has plans to plant more in the years ahead. It is expected that 18 per cent of the countryside will be covered by 2046, cloaking the slopes of hills, rooted in what once was bogland.

As we clamber, we talk. He tells me that he spent his youth and childhood in the Liberties, a working-class area in Dublin. Nor – unlike the rest of the Chieftains – did he listen to traditional music much when he was young. It was only when he began working at a printing firm that he started to take an interest in traditional melodies and songs, going to music sessions and singing the ballads and tunes he heard there, especially in the west of Ireland, the counties of Clare, Galway, Sligo and Roscommon, learning to play the bodhrán after he heard it on the radio. For all that he has travelled and played with the likes of Mick Jagger, Van

Morrison and Ry Cooder, it is still this enthusiasm that motivates him today. The love of the local can be seen in the Dublin cap perched on his head. The characters he talks about are not the superstars he may have met in the Californian version of Hollywood, or in any of the world's larger cities, but those he has encountered in the smaller Irish Hollywood, with only a hundred or so citizens within its boundaries, 500 or so scattered around. One can see that many of these people have a similar regard for him. There is real warmth in the way those living locally greet him. The schoolteacher called Sennan whom he meets in the local pub. The woman who has been working since 1983 in a pub in Greenwich Village, New York, returning for a few weeks' holiday to her native village.

Most of all, however, it is the depth of knowledge he possesses about the people of Hollywood and around. He is familiar with the type of local directory that I used to come across in my own home district. There, they were listed by nicknames. In this small community, it was the ancestry – even in terms of ownership – that was important, the way the farms and smallholdings nearby had shifted proprietors over the decades, one family giving way to another over the years. In this way, they resembled the books of croft history published by Bill Lawson and the organisation Seallam in the Isle of Harris, which tabled the ownership of land of many of the villages in the Western Isles.

It was mainly three men who took up much of Kevin's conversation: 'The Yank' McGuire given that name, someone recalled, not because he had travelled there but because he had talked incessantly about the country in primary school; the individual known as 'Cocker' Twomey, and finally Pat 'the Brute' Healy, an individual granted that name not because of any cruelty that was in his character but instead because he would head out to work even in what the Irish term a 'brute' of a day. For all the mildness of his personality, this man was able to work.

Kevin was taking me to a place which illustrated this aspect of Pat Healy's personality. This was not far from the crest of Lugglass Upper, looking down on Hollywood. It was the peat bog where he had worked for years, obsessively cutting peats in the way that some of those from my own village would have done, storing up fuel in anticipation of some endless winter he visualised way into the future, a little Ice Age like the one that froze the Gunnister Man to death. He would carve and hurl the dark stuff for days on end, standing on the bank, creating stacks which he covered up with the sacking he had taken home after working many years before on the Turlough Hill generating plant in the Wicklow Gap, the sacking used for the explosives he had taken home. The sacks had crumbled into dust over the last few years, cloaked and covered by heather grown so tall and spindly that it looked now like small bushes.

It wasn't all that had disappeared since 1995, the year Pat Healy died. At one time Pat used to have a bothy – or 'boghouse' – on the moor hidden among the banks. It has just about disappeared now; the only evidence of where it used to be a rusty tin-can peering above heather, a chimney for the fire where Pat – and occasionally Kevin – might linger for a few moments while rain and wind speared the summit of the hill. There was one occasion Kevin spoke about, when a rat scurried out, scampering up a visitor called Mick Fenton's arm. The guest made to kill it, but Pat restrained Mick's blow.

'Ah, leave her. She has a family.'

That same affection for nature, one that belied his nickname of 'Brute', could be seen too in the way he allowed birds to nest inside the thatch of his home, one that like his peat stack employed the same sacking he had taken from his work in Turlough Hill. This time, it was employed for the roof underlay. He might stand for hours above the half-door of his home, now crumbling too, watching them flit in and out around his head. The half-door where he used to lean is still in existence, used now on the music house in the village.

It is not the only memorial that features a legendary peat cutter in Ireland. Another exists near the entrance to Bellaghy Bawn in the town of Bellaghy in Northern Ireland. Unusual in the way it commemorates a poem rather than a poet, the central figure is Seamus Heaney's grandfather, who features in the work. Cast in bronze by the Scottish sculptor David Annand, the figure is hewn from the turf or peat referred to in the work's final stanza. Crouching, his extended hand holds the familiar shape of the usual Irish peat blade, the slane. On its cutting edge, there is a line from 'Digging', a reminder of Heaney's pride in having a relative who could hew out more peat from Toner's Bog than anybody else in the vicinity.

This pride is not something strange to me. I recall it from my own childhood and youth, the nod and acknowledgement of someone who succeeded in putting all the skill and experience instilled in joint and sinew into the task of cutting peat. '*Tha e math air an tairsgeir*' ['He's good on the peat blade'], they would declare, sometimes grudgingly, as they watched a master at his labours, levering fuel from the bank. It was not, however, simply a test of strength, as an old Faroese proverb underlines: *Veikur madur brytur hakan* – 'the strong man breaks the spade'. This was said when there was too much reliance on muscle and brawn, not enough on skill. One would imagine this was a universal observation, an insight of which people were aware from the Faroe Islands to the southwest of Ireland and beyond.

In Ireland, they even made national celebrities of the individuals who were most accomplished at this task, reporting their feats of strength in the national press and elsewhere. In the thirties and early forties, the Turf Development Board regularly held national competitions every June in order to encourage what they felt was The Dying Art of Turf Cutting and, indeed, in the decade before the Second World War, a similar decline in burning the fuel. They might use the *wing sleán* (slane) – the traditional right-angled tool in Ireland not dissimilar to the *tairsgeir* – in events that occurred in peat-cutting places like Allenwood or Monstarevin. In the last location in 1936, an official is portrayed in a photograph as trying to keep spectators from the edge, where they peer over like football supporters. In the caption on the Bord na Móna Facebook page that features this illustration, we are told: 'Supporters of different teams got very excited during the event. Some people were thrown out by the Gardai (police).'

Later, the caption goes on to point out the difficulties that might be caused by 'throwing' an individual out of a bog. One can only conclude that they must have used an individual skilled in throwing peats long distance for this task.

There were also breast-sleán (described overleaf) events, which may or may not have generated their own hysteria, one perhaps whipped up by the fact that board employees obtained a few hours off midweek to attend the event. In this case, each team had a cutter, catcher and barrow man; the barrow men and catchers on top of the bank. Some individuals cut so fast they needed two catchers and two barrow men, one of whom often cut the 'shelf' where the cutter would stand. It was rumoured that in some instances, the individual responsible for cutting the peat might have seven separate turves in flight before any of them touched the ground. Men like these were figures of myth and legend, the kind the Irish singer Luka Bloom immortalised in his song 'Bogman', describing a man for whom this was his natural environment.

So many people look at the bog
As a place that just lies dead
Nothing to do for the body
Nothing to do for the head
Take me where the heather and the moss grows

Yet across northern Europe, there were considerable differences in the tools employed by these heroes. In the book *The Spade in Northern and Atlantic Europe* edited by

Alan Gailey and Alexander Fenton (Ulster Folk Museum), there are a range of very particular peat spades and peat blades designed for the distinctive terrain in which people cut their winter fuel. Sometimes there were differences between two communities only a few miles away from each other, as in certain districts of Norway, for reasons that nobody could understand. Sometimes they involved the sex of those who were doing the cutting; women's peat blades tended to be a little shorter. Sometimes, though, they were more odd and alien.

In Denmark, for instance, the peat blade that uncovered the Tollund Man might have been curved and possessed a handle. (Throughout northern Europe, this was most often fashioned from willow, ash or elm.) Or it could have had a fishtail-shaped blade, as these were also common in that country. Sometimes, too, a peat scythe was used. This might even have been one of the many examples of recycling that have been prevalent in rural, peat-cutting areas for centuries. With the broken handle of a scythe rendering it unsuitable for cutting a swathe through long grass or a field of oats, it might be better utilised for sliding through the depths of peat, plying it out of its bed and laying it on the surface of the heather. There was an implement resembling an Archimedes screw in Jutland, where the terrain was soft and wet. Hand turned within a tube, it would extract some of the moisture as it was hauled from the depths. In some parts of Shetland, where there was the opposite problem and the peat was crumbling and dry, they used a delving spade. In other locations, such as Ireland or the Yorkshire moors, a breast-spade – or 'breast-sleán' – was employed. This was when a man would cut the fuel from the side of the bank. A variation of this was the flaughter-spade or breast-plough. Despite its name, this was pushed down from someone's thighs, a crossed handle at its top. Often, the peat obtained this way did not have the prime purpose of generating heat for the household. Instead, it was stacked, dried and quickly burned; the ash used as fertiliser for the fields. Again, this is a way of using peat that is not that far removed from my own experience. There were mornings when it was my task to remove ash from below the grate in the fireplace of our sitting room or the stove in the kitchen. There were mornings when it appeared light snowfall accompanied my progress through these rooms. White ash settled on the fireside chairs and kitchen table. It swirled when I stepped outside, making my way towards the manure heap outside, a blizzard flurrying even on the most still and warm of days. Finally, I would empty the pail of the little ash it now contained, sprinkling it over a mound of cow dung and damp straw.

Some of the tools for cutting peat that are described in the book, however, are not so familiar. There is one called a 'ripper' with a two-sided foot-rest; another termed the *haki* in Faroe for cutting and paring turf. The most common tool there for peat cutting and other tasks is the *tovskeri*. Similar to a cricket bat, like some found in parts of England, it must have its limitations as it relies on the strength of a person's arms only. There are various tools used in Aberdeenshire in Scotland, a veritable arsenal to ensure that the peat or turf is cut. They include such idiosyncratic and precise items as the bullin spade, turned bullin spade, stamp spade (one side), two breast spades, two tirrin spades, equipment to accomplish every task from the cutting of the heather on the surface to the digging out of the deepest, darkest fuel.

One further suggestion as a means of cutting peat has come from the pen of Irish writer Myles na Gopaleen, also known as Flann O'Brien, also known as Brian O'Nolan. In his column in the *Irish Times* in which he chronicled the excesses of the Irish Civil Service, a body for which he worked, he included a plan for the national railway network. Henceforth, he declared, it would only run across the country's bogland. The engine would be complete with a peat extractor and a drying machine, scooping up turf as it rattled along, applying a gentle heat to each sod of turf before using it as fuel. Myles recognised there might be some dangers in this approach, especially in the way that certain locations in the Irish moorland were frequently used for the production of *poitín* (illegal alcohol). It was as a result of this that anti-flash protection gear was to be provided for the engine crew in case deposits were encountered by accident.*

There are a variety of reasons why peat inspires such wonderful moments of madness in Ireland. It is a country that is dominated by bog, possessing two principal types. The first of these is the lowland blanket bog found across much of the country, in places like Galway, Mayo, Cork, Kerry, Donegal. Based on a layer of acidic rock, they experience high levels of precipitation and provide a basis for human settlement. In other areas, there is the upland blanket bog, which occurs between 150 and 350 metres in the west of Ireland, becoming mountain blanket bog above 300 metres. Bilberry and crowberry thrive in these areas, the latter being an alpine species.

But almost as much as bog dominates the country's landscape, it also plays a large role in its industrial and economic history. One can see the beginnings of this in the late nineteenth century. This is especially the case in County Offaly, a location I circled in my car in the late spring of 2016, driving through its nature reserve and wasteland; each mile bringing its own reminder that, unlike oil or coal which is – in the main – extracted from deep within the earth, in the case of peat, the landscape is altered for ever. In the 1890s, for instance, Kieran Farrelly started a peat-harvesting business at Turraun Bog. Some ten years later, he was the proud owner of 100 hectares (240 acres) of bogland under development and

* It is not the only Myles story that features turf. On another occasion, Myles has Sherlock Holmes consulted by a client claiming to be the Irish legendary hero Oisín, looking for the Fianna. Recognising his arch enemy Moriarty through his disguise, Sherlock sends the villain off in a turf cart. We discover the two of them incarcerated and institutionalised by the story's end.

had built a factory to process the peat. However, his success did not continue for long. After a flood destroyed his factory in 1903, in itself a dramatic reminder of what occurs when layers of peat are stripped from an area, Farrelly was forced to emigrate to America. The Turraun Peat Company was bought by a Welshman, Sir John Purser Griffith, in 1910. A highly influential individual in this industry for the remainder of his life, Griffith drained Turraun Bog. Later, in the twenties, he built a peat-operated power station that produced 4,000 tons of sod turf each year, bringing light, perhaps, to rural areas, a process that Irish poet Maurice Riordan describes in magical terms when electricity – much later in 1956 – was brought to his home in rural County Cork:

> the lights of Piccadilly
> were swaying among the lamps of fuchsia*

Much of the turf, however, was transported to Dublin via the Grand Canal that crosses the country, bringing heat to the homes in the city. In 1936, the Turf Development Board – a body set up in 1933 – purchased the company. This was another sign of how there was political support for the enterprise, especially from Frank Aiken, a former IRA man from Armagh in the north of Ireland and the Minister of Defence in the Fianna Fáil Government. A restless and inventive individual, who among his other activities took out patents for the invention of a new kind of turf stove, beehive, air shelter, electric cooker and even a spring heel for a shoe, Aiken wanted to create an industry based largely around the practice of cutting turf by hand, a practice that was dying out in the early thirties, decreasing from 6 million tons in the

*The more house-proud in rural communities did not universally welcome electricity. Far more than the old-fashioned Tilley or oil lamp, the bright gleam of electricity showed up the layer of light peat ash that frequently fell upon the furniture and floor. Before that time, it was undoubtedly a case of 'what the eye cannot see'...

early twenties to just 3.3 million in 1933. As a way of preventing emigration and combating poverty in rural areas, he planned to ensure what was termed a fair 'bogside price' and quality standards to the product, determining, for instance, the density and moisture levels of peat that was on sale, making certain, too, that there was a proper transport system to bring turf to the towns and cities where the fuel would be burnt. This involved not only making certain the lorries that carried the peat were fit for purpose, but also that the network of roads that crisscrossed Ireland were. While the two men probably did not know each other, together with his colleague C. S. Andrews he had much in common with Tom Johnston, the Secretary of State for Scotland during the Second World War and the Labour government of Clement Attlee. There was a similar energy and commitment to the two of them, particularly regarding their attitude to the neglected, rural areas of their respective countries. It was summed up by his statement when Bord na Móna was created in 1946, as a result of the Turf Development Act, declaring it was 'a crusade rather than a commercial project'. There is something of this attitude, too, in the combined words of Aiken and C.S. 'Todd' Andrews – Andrews being managing director in the early days of the industry – when they decided the term 'turf' was to be rejected, declaring it to have inferences of 'inertia and ignorance'. The Irish *móin*, genitive *móna* used in the new title, like the equivalent of 'bog' (*portach*) have no such disparaging connotations.

In his autobiography, *Man of No Property*, C. S. Andrews goes further than this. A 'freedom fighter' born the son of a relatively prosperous owner of a dairy business in a poor area of Dublin, he tells of his first visit to a peat-cutting area in rural Ireland. He had no awareness of what they were doing there, only knowing:

> 'that the word "bog" or any phrase containing it, had become the symbol of poverty and backwardness. The "bog-trotter"

> was the Irish archetype of ignorance and illiteracy. The bog men were the descendants of the Fir Bolgs, the bog itself in the Irish mind was a symbol of barrenness. A large area of the Irish "race consciousness" was overlaid by bog.'

It is an attitude that is echoed throughout these islands, still present in the term 'culchies' applied to country people by the city-dwellers in Ireland or words like 'teuchters', 'hicks', 'bumpkins', 'yokels' or 'hayseeds' that are hurled in the direction of rural people elsewhere in the world. One can see it in the attitudes of those dwelling in the tenements of Dublin in the twenties, where 'Captain' Jack Boyle in Sean O'Casey's play *Juno and the Paycock* consoles himself with the thought that: 'We're Dublin men, an' not boys that's only afther comin' up from the bog o' Allen.'

In the early years working in the Turf Development Board, Bord na Móna, Andrews was always conscious of the effects of this. The opposition party, Cumann na nGaedheal, which later transformed itself into Fine Gael, had little sympathy for the entire notion of creating an industry out of the boglands of Ireland, arguing against this on every occasion. In this, they were often supported by the coal merchants of Ireland, who resented the subsidies given to their native fuel, especially in the early days of the peat industry. Andrews did not always either receive support from Fianna Fáil, the government party, noting bitterly that the criticism against these developments were often at their most intense from those 'born on a bog' and valuing the coal fires of their new home. They were, after all, more aware than most that in terms of producing heat, turf was much less effective than either coal or wood. It also involved considerably more time, patience, labour and dry weather. As far as organising a way of transporting it elsewhere, they believed it was 'better to burn the bog where it was born'.

It was the Second World War – or what is called in that country the 'Emergency' – that changed Ireland's attitude to its greatest natural resource. Raw materials were scarce, especially since, as Andrews argues, the country's 'economic connections with the outside world were largely second hand'. Even tea arrived upon its shores through its closest neighbour. Coal was one of the supplies most affected. This may have been partly caused by the danger to colliers crossing the Irish Sea from England. Andrews ascribes it to something deeper. There are 'three things to beware of', he writes: 'the horns of a bull, the hooves of a horse, and the smile of an Englishman'. Whether the lack of coal was the result of either accident or animosity, there is little doubt that, after the spring of 1941, there was a shortage of fuel to heat Irish homes.

It was a situation that led to a change of attitude to peat across Ireland. The entire county council organisation was put in charge of turf production across the country, linking up with the Office of Public Works to ensure the fuel could be transported across Ireland through its road, rail and canal network. The Land Commission and the Special Employment Schemes Office were enlisted to open up and obtain new bogs by private individuals. Rationing and price controls were introduced in the six counties that made up the eastern edge of the country, where turf was not available. In places like Phoenix Park in Dublin, people would arrive with ration cards in hand, asking for their household's allocation. To many, this was a depressing time. Women would often (rightly) wave the sodden turves of peat they had been allocated in the face of both official and husband alike, complaining about the nature of the fuel they had been given. It was an experience which Andrews, one of those in charge of Bord na Móna, even experienced in his own home, his wife appearing before him at his sitting-room fire one evening brandishing a clump of peat.

'How do you expect me to burn the likes of that?' she asked.

The dependence on 'home-grown' fuel, however, also chimed with attitudes that were in existence in Ireland long before this time. There is apparently an old Irish proverb that declares 'He who has water and peat on his own farm has the world his own way', a definition of self-reliance and personal independence if there ever was one. There are also the words of Jonathan Swift at the turn of the eighteenth century. The author of *Gulliver's Travels* made his own pronouncements on the topic, noting in his 'Proposal for the Universal Use of Irish Manufacture' that the men and women wished 'never to appear with one single shred that comes from England'. He also predicted that Ireland would 'never be happy until laws are made for burning everything that comes from England except their people and their coals'. For all the regrettable absence of coal from the grates of their household fires, the Emergency provided them with the justification for doing just that. It became part of their nation's wartime leader Éamon de Valera's basic economic philosophy to ensure the self-sufficiency of the country in virtually all goods, ranging from sugar (and the sugar beet industry) to shoelaces. Clearly this might mean that the lives of the people would be modest in terms of their income, but he saw all this as part of his utopian vision for the country, one he declared in a rather pious and dewy-eyed radio broadcast in 1943:

> The ideal Ireland that we would have, the Ireland that we dreamed of, would be the home of a people who valued material wealth only as a basis for right living, of a people who, satisfied with frugal comfort, devoted their leisure to the things of the spirit – a land whose countryside would be bright with cosy homesteads, whose fields and villages would be joyous with the sounds of industry, with the romping of sturdy children, the contest of athletic youths and the laughter of happy maidens, whose firesides would be forums for the wisdom of serene old age.

The industry made its own contribution to de Valera's dream of a countryside 'bright with cosy homesteads'. New houses were built in many areas where people had always emigrated from before. In a venture that reminded me of the place names of Newfoundland I used to pretend to glimpse from my home in Lewis, advertisements appeared in Ireland's main newspapers. They encouraged some to reverse a process that had occurred through previous generations, leaving Dublin for small communities that might be made 'joyous with the sounds of industry'. Designed by Frank Gibney, a man without formal planning or architectural training, houses were built for the workforce in places like Kilcormac in County Offaly, Rochfortbridge in Westmeath and Coill Dubh in County Kildare. Coupled with providing houses for those employed on the turf, it was also in some ways an attempt to reverse history, to go 'back to the future' with much of the energy of Marty McFly. If this idea appears somewhat fanciful, it might be worthwhile considering this from Bord na Móna's website. Summing up his architectural work and the role of Todd Andrews within it, the website declares:

> While Gibney's work for Local Authorities was of the highest order, Bord na Móna was a unique client. It is said that the three characteristics for good civic design are a skilled architect, an understanding citizenry and a powerful and benign king. This combination gave us Leningrad, Central Paris, and Brasilia. In Todd Andrews, Gibney found his Sun King and however they may have felt about it at the time, certainly the love which the inhabitants of these villages have subsequently lavished on their environment indicates their pride and satisfaction.

The vision of 'national purity' conveyed by both De Valera's words and these words has, of course, much in common with the politics of the time across Europe, during the thirties and forties. De Valera's version – like that of Dollfuss in Austria, Pilsudski in Poland, Salazar in Portugal – was in relative terms more benign than some of its counterparts elsewhere in the continent but it still consisted essentially of nostalgia for a country and era that was gone, if indeed it ever existed at all. It would also today be condemned as the worst kind of 'austerity politics' with 'the laughter of happy maidens' and the existence of 'serene old age' becoming notably in short supply after a time of 'frugal comfort'.

The headquarters of Bord na Móna even today have something of the atmosphere of that era, one in which residential camps for labourers were set up in many areas where peat could be found in Ireland. (Fourteen were to be found in the Kildare area, each with a capacity for 500 workers.*) Situated in the former garrison town of Newbridge in County Kildare, a short distance away from a modern shopping centre with Debenhams, River Island and Marks & Spencer, the organisation's main offices are to be found in what was once a cavalry barracks, occupied by the British Army until 1922. Behind the imposing pillars of its entrance, it is almost possible to see the ghost of Lord Cardigan, leader of the Charge of the Light Brigade in the Crimean War in 1854, parading with other ranks on his steed. The sweep of its grounds and the array of red-brick buildings within its walls have that aura of a long-established empire, albeit one now that has shrunk and is largely based on the peat bogs of Ireland.

It was within its walls that I encountered one of Bord na Móna's employees, Tony McKenna. A pleasant, reserved man with silver hair, he and I spent much of our time talking about all we had in common. Both of us were brought up in the bogs. As a native of Newbridge, he had spent much of his youth cutting peats. We had both observed how the efforts to transform the moor near our homes into fields had been failures. Too much fertiliser had been required to create apportionments of 'improved' land; within a few years

* In May 1944, some surprising recruits joined the workforce. Some 80 German internees imprisoned at the Curragh Camp during the Emergency were taken on as part of the Kildare Scheme. Their work proved 'satisfactory'.

more rush and reed had grown than grassland. We had also seen how – as Kevin Conneff and I had observed – the moor had been transformed into a site for forestry or endless rows of pylons to generate electricity in a way far from the one Frank Aiken and C. S. Andrews might have envisaged when they first contemplated the moorland's capacity for generating heat and light. We spoke, too, about the battles that were occurring on some peat bogs, where local residents continued to cut their winter's fuel for all that there were laws and directives urging that practice's end. It was, we both agreed, something that these two individuals who had helped set up the organisation could never have foreseen.

There were other things that had altered since the days of these two men. In the email correspondence that preceded my visit – not, I should point out, written by Tony – there was little evidence of the (extremely eloquent) bullishness to be found in the biography by C.S. Andrews. In his work, he complains that Bord na Móna is not given its rightful place among other national organisations such as Aer Lingus or Córas Iompair Éireann, the transport network. Instead, there was a reluctance to speak, monosyllabic responses, a grudged 'OK' or two. There was also, as a result of environmental pressure, much less self-confidence on display than there had been some 60 years before.

Instead, this quiet, modest man spoke of other approaches to the delivery of energy – how they had embedded pylons for windfarm schemes, how they were encouraging biodiversity in their territory, how they were using sawdust, bark from felled trees and those too small for milling, plus other forms of biomass generating power in their Edenberry site, how employment levels and the use of machinery had declined in recent years. One could imagine the ghosts of Aiken and Matthews, even those of Lord Cardigan, treading softly and listening to Tony McKenna's words with an air of startled surprise.

CHAPTER SEVEN

Svörður (eastern Iceland) – Turf, Peat

My great-uncle Allan was the one man I associated more than any other with the moorland during my childhood. During the last days of his life, however, he spent all his days lying in his bed, his old and grizzled face looking up briefly from his pillow when I entered the room.

'Hello, Dan,' he would say, using one of the diminutives that were sometimes employed in the village for those with my forename.*

And then the conversation would slowly come to a halt. With the way his breath rattled and wheezed, he had little strength for conversation. Instead, my legs dangling from the edge of the bedside chair, I would sit and scan the names of the many books stacked beside him. Their spines bore titles like *The Empty Land*, *To the Far Blue Mountains* and *Where the Long Grass Blows*; their authors had names like Max Brand, Louis L'Amour and W. C. Tuttle. Within their pages, characters called Hondo, Edge or Sudden stalked the open prairie or desert, rode across the Rocky Mountains with shotguns or Colt 45s fixed in their holsters, intent on mayhem or revenge for whatever rough hand fate or the local ranch-owner had dealt out.

The young men of the village used to go down to my great-uncle Allan's bedroom in his home in South Dell in the Isle of Lewis to read these stories to him. He lacked the energy to turn the pages himself, forced to listen instead to their Hebridean voices struggling with such alien terms as

* Another was *Doilidh* or 'Dolly'. I was always spared this. For such small mercies …

Appaloosa, renegade, showdown at the OK Corral. No doubt there were times when he must have sniggered quietly as they tried to pronounce the names of the Native American tribes – *Comanche, Mohican, Mohawk* – or the odd places, like Tombstone, El Paso or Shiloh, where cowboys and rustlers lived.

Much of Allan's life had been spent out on the Lewis moor – the largely flat, brown 'prairie' that stretched between Ness and Tolsta at the tip of the island. He would scramble over peat bank and tussock, skirt loch and bog as he journeyed across land that was dangerous and deceptively difficult to cross. Mist would sometimes descend on it, moist earth would suck at a man's feet, more water than land even when appearance might suggest otherwise. Its heather-covered ridges can give way suddenly to uncanny black.

These walks had not been purposeless rambling either. Allan had been gifted with an almost encyclopaedic knowledge of the sheep that grazed there for much of the year. He could recite the colours, for instance, that daubed and banded the backs of different flocks, identifying their owners, even if they lived on the other side of the island. He was able, too, to tell which crofter they belonged to by the way their ears were cut and marked. '*Toll 's beum air cluas taisgeil*,' he might say, part of the litany of cuts, nicks and holes that showed on which croft they had been born, the name of the man who had taken his blade to that part of their anatomy to claim ownership of a particular ewe, lamb or wedder.

Later, he might limp with that knowledge to a crofter's house, telling them, perhaps, that one of their ewes had gone down in a peat bog near Loch Dìobadal or far out near Mùirneag or that a flock might have wandered far from the village, grazing on the outskirts of North Tolsta on the island's eastern side. 'You'll need to go out and bring them in,' he would advise the owner.

This usefulness would have surprised those who had seen him in the early days of the Great War. He had been taken home from the mainland on the passenger boat the *Sheila*, at that time with one whole side of his body paralysed. He was unable to use his left hand or leg; his smile touched only half his lips, the other part curling in a permanent grimace. Unlike his brother, John, killed in France at the age of 22 on 22 May 1915, Allan's injuries had not, however, been caused by a bullet. Instead he was the victim of rheumatic fever, caught – according to family folklore – when he slept, wrapped in a damp blanket, in a boarding house in Liverpool. He had never even stepped on board the ship where he was due to serve for the duration of the conflict.

Yet he staggered on, living with his sister and her husband in the blackhouse where they raised their two children, and employing that strong right hand of his to help them on their croft. It was, perhaps, for this reason that he was always 'out': to give them peace and privacy to be with one another. To be without constant reminders of how much of his manhood had been lost through his injury. To be on his own, far from other people's eyes and pity. In order to keep his dignity, there was always one last act he performed before heading out the door each day: using his right hand to place his left one in his jacket pocket, he would try and conceal his disability from view.

He continued these moor walks of his until he was in his seventies, gaining an expertise which none of his people own today. In this way, he came to know the names of plants and landmarks on the moor, each sign and omen of the coming weather. It was for this reason, too, that I took pride in him. My primary schoolteacher used to ask about him regularly.

'How's Allan?'

As best as I could, I would give an answer.

'The best shepherd in the district. Always on the go.'

It was a freedom he enjoyed until the day he slipped from the stones that acted as steps on the outside wall of a blackhouse, helping others thatch its roof. His other leg was shattered that afternoon. It was an injury that had forced him to lie sprawled in his bed for the remainder of his life, listening to the young men reading aloud the western books they had brought him, these tales of Hondo, Sudden, Edge – all these dark and brutal tales of men who lived and travelled alone.

His western books were not the only tales of men who lived and travelled alone I heard in or around my home in Lewis. There was occasional talk of eccentrics and hermits, the mentally ill and those who were unable to settle among others after a time of war. One of these individuals is written about in the book *Tales and Traditions of the Lews* by Dr Donald MacDonald of Gisla. He draws attention to a man he calls Dòmhnall Aonghas Buachaille (Donald Angus the Shepherd) who lived on the moor in Uig on the west side of Lewis. Apparently, this legendary figure wrote:

Though I now live on Bastair's edge,
I much prefer it to serving the King
Where with trigger tight against finger,
A hail of lead slashed wounds upon my flesh.

I also listened, both in school and in the community, to many stories about a legendary murderer, Mac an t-Srònaich, who was said to have roamed the moors of Lewis during the middle of the nineteenth century. He was reputedly a real individual, said to be Alexander Stronach, son of an innkeeper in Garve at the eastern edge of Rosshire, and grandson of Alexander Stronach, minister in Lochbroom in the west. The only reference to him in historical documents is a document entitled: 'Proc. Fiscal v Bodach no Mondach or Fantom. A Moor Stalker.'

These words seem to be a reasonable summary of him too. When Derek Murray, a Gaelic DJ and broadcaster, and I spoke about this figure whom we had first heard about in

Cross Primary School many years before, there was little doubt in our minds that this is what he consisted of, a 'phantom', an elusive figure whom it is difficult to grapple with even today. His existence (or not) had been a source of such fascination to Derek that, a few years before, he had even written about a play about the myth of Mac an t-Srònaich. It was a legend that was so powerful that, as the Hebridean writer James Shaw Grant pointed out, his name was 'being used to frighten naughty children in Canadian families three or four generations removed from the Hebrides'.

Both Derek and I could recall a little of that; he was the 'bogeyman' whose name was whispered sometimes to prevent us stepping out far into the moor, keeping us away from a loch brimful with water or the confinement of a bog. I recall one elderly woman in the village who swore – somewhat improbably – to have seen him coming towards her, jumping into a peat bank to evade his grasp. There was another account from the village of Tolsta across the moor, when 'a young woman was chased' until 'she came in sight of the village houses'. It was then 'she started to shout and her pursuer stopped chasing'.

Sometimes there seems to have been an element of the morality tale to the legends connected with him, as there is in many other stories about the Hebridean moor. There is one, for instance, set in Gisla in Uig in the south-west of Lewis. This tells the story of a man called Macaulay who killed an Irish pedlar and harpist who was in the area, burying him in a shallow grave. The reason for his act is the man's beautiful wife, whom he takes home with him, unaware that she is pregnant. A short time later, a child is born bearing the features of his victim. He is haunted by his constant awareness that the forebears of the one he tries to pretend is his firstborn are known as *Sliochd a' Chlàrsair* (the Harper's People). At other times in these stories, the 'ghost' is a much more traditional form, stalking either the murderer or the community like a shadow that constantly follows them around.

A different kind of morality is served up in the Mac an t-Srònaich stories with none-too-subtle warnings about the perils of pride or vanity. One involves others in his family home in Garve, where his father was at one time the innkeeper.* Apparently, according to the North Tolsta historical website, 'he had a sister who was friendly with a young girl who was staying with them'.

> The two girls always shared a bed – there were no single or doubles in those days. These people who stayed with those in the Stronach house were fairly well off, and the girl had a gold chain around her neck. Mac an t-Sronaich had his eye on it, and he meant to get it somehow. He thought he would go into her room while she was sleeping and snatch it, and no one would know anything about it.
>
> The girls, being girls, were silly, and Mac an t-Sronaich's sister asked the other one if she could wear the gold chain around her neck for the night, as she herself didn't have anything like that. The visitor let her have the chain for the night. Mac an t-Sronaich chose that night to go into their room. When he grabbed the chain, the girl awoke and cried out, and he killed her there and then, in case he was caught. Of course, it was his sister, and when he found out in the morning, he ran off and went into hiding, coming to Lewis and living for the rest of his time as an outlaw.

Clearly, his sister is being punished for her vanity on this occasion, wanting a gold chain few native Highlanders could afford.† There were other occasions when the 'murderer'

* There may be some significance in this. 'The Dark Mile' outside Garve was a place of rather dubious reputation, known for robberies and attacks on those travelling through the area.
† In some versions of the story, it is a string of beads.

punished people for the sin of pride. This is particularly the case when the executioner taunts him at the gallows, an event that allegedly took place at Gallows Hill in Stornoway. The hangman points out that he only killed 19 people, not an even figure of 20. An instant later Mac an t-Sròinaich takes the tankard he had been given to ease his thirst and uses it to kill the man employed to execute him.

There is no record of this event and precious little else about Mac an t-Sròinaich's life. Even the portrayal of him as a murderer is subject to dispute. He was more likely to be a fugitive from mainland justice who believed that when he reached the Isle of Lewis he would be free from persecution. Instead, as is the case even now when people attempt to escape their personal problems and head to rural areas, legends tend to multiply about them. No one trusted the outsider who clearly had his own dark story to tell. They ascribed murderous deeds and unsolved crimes to him. Natural deaths on the moor occurred because he might be in the vicinity; murder even before his time retrospectively declared the work of his hand. It was for this reason that he became more and more desperate in his behaviour, attacking other poor people in the community who might be in the possession of food or a little money.* He may even have been unable to communicate with them, such Gaelic as he had (if he possessed any) an inexplicable mainland dialect to their ears. It was all a mixture of elements that probably added to his air of desperation.

Sitting with me in the BBC canteen, Derek spoke with an air of solemnity, something that contrasted with his usual

* Oddly he seems to have obtained food from the more prosperous. He was 'one of them' and seems to have been fed sometimes by ministers' families and those who owned businesses in Stornoway. Clearly they knew something of his background and situation, as opposed to the Gaelic speakers in the country who saw him only as a threat.

mischievous, extrovert behaviour, the one that was familiar to me when he was on air.

'He was a *truaghan*. A poor soul. He was probably extremely hungry and isolated. There is little on the Lewis moor to eat, especially in winter. Especially when you're the son of pretty well-off people and have no experience of foraging to keep yourself alive. Especially when you're an outsider in an area where people have good reasons to fear outsiders.'

It is this last aspect that is possibly least understood about tales like Mac an t-Srònaich. The people of isolated districts in Lewis had good reasons to be fearful, especially of outsiders, at the beginning of the nineteenth century. They lived in the period of the Highland Clearances when there were tales of villages being emptied, homes evacuated of those whose families had lived in them for centuries. Some of the leading agents in the destruction of that way of life were those who were 'Highland gentlemen', from a similar stock as the legendary murderer. As James Shaw Grant in a letter to the *Herald* notes:

> His significance, which has not so far been adequately researched, or even recognised, arises from the fact that he was operating at the interface between the English island 'establishment' of the time and the Gaelic majority.
>
> On one side of the line his identity was known and he was given succour because of his connections. On the other side of the line he was a danger to wayfarers. He lived rough, stole and resorted to violence when he needed food.

James Shaw Grant goes on to write about how these two cultures sparked off each other, pointing out that at this time, in the 1830s, he had little doubt that the 'English' island establishment was the superior one. The reasons he gives are interesting. They hinge on his belief that the

'English' establishment was bilingual, enriching its native culture from another source, while the Gaelic majority were monolingual, drawing on a culture in decline. Paradoxically, however, with English becoming the prevailing language of the Highlands and Islands, he goes on to claim that later it is those who speak Gaelic who have the advantage, gaining access to two different tongues and ways of thought.

As I look back at my own youth and childhood, there is some substance to his view. From the late sixties onward, I spent much of my childhood and youth watching television. There was little there that reflected the world in the vicinity of our peat stack. One exception was *Sutherland's Law*, which starred Iain Cuthbertson as a Procurator Fiscal* in a fictional Oban, called Glendoran. It was set in a landscape of fishing boats and moorland; its swirl of scenery accompanied by a soundtrack that featured Hamish MacCunn's 'The Land of the Mountain and the Flood'. It also possessed its share of clichés and stereotypes. Some even came from the age to which Mac an t-Srònaich belonged, with city visitors often unwelcome among its glens and mountains. There was even the obligatory Glasgow villain who sidled up to Cuthbertson's considerable bulk, telling the lawman: 'I wouldnae like to see a nice old guy like you making a fool of yourself. I don't doubt for a minute you're the terror of the local poachers and the guys who throw toffee papers away on the street. Do you not think you're a wee bit out of your class?'

And there was literature, too, some of which featured moorland not that dissimilar to the empty miles that stretched out before our sitting-room window. (The house pointed in that direction rather than the infinitely more scenic view of sea and beach on the other side of our home. The reason? It was easier to spot the arrival of both

* A Scottish legal officer responsible for criminal prosecutions.

bus and mobile shop – bearing bread, meat and fish – when we gazed out there.) Like most of my classmates, I also encountered Enid Blyton and the world of books. The Famous Five seemed to be continuously encountering criminals and eccentric individuals on landscape not unlike our own. *Five Go to Mystery Moor*, the title of one proclaimed; *Five Have a Moorland Picnic* the name of a jigsaw that featured a similar scene. 'We should be there about half past five. Look, you can see the moorland in the distance now – all ablaze,' a character proclaimed. We are told 'it looked wild and lonely and beautiful, blazing with heather, and shading off into a purple blue in the distance'. There was little doubt that to a child living an urban existence, the use of such a setting might have provided an exotic element to the plot, allowing the author to create a landscape far removed from the one where parents made decisions for them, where youngsters could face their own difficulties and dangers by themselves. For me, however…?

By the age I was encountering Enid Blyton, about 11 or 12 years old, I was already going out to help my father and others with the peat, filling up a wheelbarrow, emptying it into a stack. The moorland was not always a place where I escaped parental authority, but sometimes where I experienced it at its most intense, provided with continual instructions about what I should do and when I should do it. It was probably for similar reasons that *Wuthering Heights* had little appeal. In that work, the moor constitutes freedom. It is where Catherine and Heathcliff escape the restrictions of the Hindley household, 'one of their chief amusements to run away to the moors in the morning and to remain there all day'. We are informed by Catherine, too, about Heathcliff's similarities to the moorland around him. He is 'an unreclaimed creature, without refinement, without cultivation; an arid wilderness of furze and whinstone'. There was an occasional man in our community that could have fitted similar descriptions. I must confess they were without much fascination for me. One can wonder if, with their cloth caps and wellington boots, they had a similar effect to that of Heathcliff on the young women in the district.

From the Gaelic stories I heard around the peat-fire flame I drew a different kind of inheritance from the moor. Some were morality tales, many similar to those told to Scottish artist Will Maclean when he and his family travelled along An Dìridh Mòr, a vast stretch of moor between Garve and Ullapool, on their way to Coigach every summer. As if the journey were not dangerous or threatening enough travelling in a Ford 8, a vehicle known for its spectacular lack of brakes, Will's father would entertain his passengers with a series of dark tales he had heard earlier in his life. As his car heaved up the slope from Aultguish Inn, an old hostelry that has stood there some 400 years, since the days when Highland drovers used to urge their cattle south to the markets in the Central Belt, he would recount a story about how some island fishermen had reached the inn after being robbed of their belongings outside Garve. A short time later, a minister arrived, a stick in his hand. It transpired that the clergyman had used this on the same thieves, obtaining the grief-stricken fishermen their money back. Chastising the men for their cowardice, he threw the money on the floor in front of them. He also used the situation as a prompt for a sermon, telling them they should trust the Lord – or at least His emissaries on Earth – more than they did at present.

And then there was another story, more grotesque, about how in the nineteenth century a coach became stranded in a snowdrift in the same area with a young pregnant woman on board. Fearing she might freeze to death, the coachman killed one of the horses, took out its intestines and wrapped her in the animal's belly. They were rescued the following day and the girl survived.

Will also pointed out that the tale possessed a Victorian ending, one that informed listeners that virtue brought its own reward. Some time later the coachman fell upon hard times and was wandering hopelessly around the streets of Glasgow. As he passed a grand house, his name was called out by the lady he had rescued. She looked after him for quite some time, until his belly was full once more, the worn soles of his shoes repaired.

And then there are those told in Shetland. Some of them were shared with me by the unmistakable form of Davy Cooper, the Shetland storyteller. A large, well-rounded man with a cap and an equally large, well-rounded dark beard, he was in Mareel, Shetland's Arts Centre in Lerwick, when I met him. It was a summer evening and the first Shetland Boat Week was taking place. We looked out the window towards the museum where the event was occurring, watching the flags flap and visitors examine the vessels on display, hauled up on the quay.

Davy comes from Mossbank, one of the villages most affected by the coming of oil to Shetland. Built upon a bleak slope of moorland, it owed much of its existence to the sea it slanted towards, either through the men employed by the Merchant Navy or through fishing. It was a precarious existence. At the end of the sixties, Mossbank as a community was fading and closing in on itself, the number of its inhabitants only around 50.

And then, in the seventies and eighties, another kind of wave rolled in, the arrival of the oil industry. The community grew in size until there were over 10,000 within its boundaries when the camps were in operation during the early seventies. Nowadays it has dwindled once again, settling at present at around 200 to 300 people.

All of this affected young Davy. The son of the janitor at the local school, a man he describes as 'a bit of a community activist', there was no doubt the first group of newcomers were extremely welcome in Mossbank. Not only were they respectful towards their hosts, they also introduced and/or revived the drama group, youth club, girl guides and other bodies, making sure there were activities for young people to attend. However, as the array of new arrivals increased, so did the tensions. Groups would besiege the local dance halls. On one side, there would be the 'sooth-moothers', those who had arrived on the ferry through the 'south mouth' of the harbour. On the other were the 'Magnies', the local lads from Shetland.

Whatever the results of the fisticuffs on a Friday night in these encounters, there was no doubt that in cultural terms there was only one winner.

'My generation abandoned its roots,' Davy declared. 'We stopped going – like so many of our forefathers – to the Merchant Navy. Both the croft and the peat-hill lost its hold on us too.'

It was perhaps in reaction to this that Davy travelled in the opposite direction. Though his usual employment was as a communications officer with Shetland Islands Council, he also, with the encouragement of both the late Lawrence Tulloch and Charlie Laurenson, became a part-time storyteller, drawing in his audience with tales that belonged, perhaps, to the islands' moors, like that of one-eyed Gibbie Law, a handsome young man from Walls on the west side of Shetland:

'He was courting a young woman from Brig of Walls, one of the nearby communities. The night he was to become betrothed to her, he had set off, as was the usual custom, with a bottle of whisky clutched in his hand to offer his future father-in-law a drink. He knew there would be a show of reluctance before every last dram was drained from the glass. He was halfway upon this journey before a man came up on his blind side. This was Simon Arthurson, his rival for the young lady's hand. A moment or two later, there was an altercation, Simon stabbing Gibbie with a quick jerk of his knife. After that, Simon panicked. He buried Gibbie in a shallow grave and rushed away to Lerwick, signing on as a crewman on a ship that was going to America and back.'

Davy paused, sipping from the cup of coffee on the table before him.

'And after that, life went on as normal. People assumed that Gibbie had suffered cold feet at the prospect of marriage, turning his good eye to look at life elsewhere. As for Simon, they knew he had been unhappy for quite some time. Perhaps he, too, had headed off elsewhere. In that small community, the thought of murder never entered their heads. It was only Gibbie's dog that was restless, pining away, looking everywhere for his master. One day, while others were digging for turf, she finally found him, scraping away the peat dust until a hand was exposed. It was then she began barking, bringing others to her side. Eventually, Gibbie was removed from his grave and brought to the kirk. He was laid out on a table there, stretched for everyone to identify him by his dead eye, and to see, too, the gash that had been made in his clothes and side.

'As in many of these stories, coincidence plays its part in what happens next. It just so happened that Simon had just returned from his ship. It just so happened, too, that people put two and two together and decided his disappearance at that time might have something to do with Gibbie's death. It was for this reason that all the men in the community were asked to go to kirk. In order not to rouse suspicion, Simon decided he had to obey the instruction. He must have gone pale, however, when he heard one of the older men in Walls coming out with the next instruction, asking all those who were in attendance there to lay their right hand upon the corpse to prove their innocence. One after another, the people stepped forward to do just this, most doing so firmly, confident that they had nothing to do with his death. Simon's fingers, however, faltered. When he brushed against his rival's skin, a single drop of blood ran from Gibbie's blind eye.'

Again Davy paused, looking out the window at a group of island dignitaries gathered on the steps of the museum.

'It was taken as evidence of his responsibility for the crime. Shortly after that, he was hanged.'

Yet it is not only moralistic murder stories like Davy Cooper's tale that are told about the moor. There are also weird and unsettling stories, the equivalents of the urban legends found within our cities. They include sightings of imaginary cars, complete with glaring headlights, speeding down roads where there were not even tracks in existence at that time, funerals occurring in empty areas of moorland, ghosts – like the ubiquitous phantom hitchhikers – that haunted the area, say, around the Ness Hall near my home. (This was built on a location they called Druim Fraoich, a 'heather ridge' that existed on the boundary between my home village of South Dell and its neighbour, North Dell. A *cailleach* – old woman – was reputed to wander there in perpetuity, or at least until a young woman died in the vicinity.) There was also a meeting with a monster one of the villagers experienced one Friday night. Not up to the standards of Nessie or Morag the Loch Morar monster, though. For all that it made the pages of the *Express*, *Mail* and my favourite boyhood comic, the *Rover and Wizard*, it was not a hugely impressive beast – barely the size and scale of the average cow.

And so it was throughout many of the communities that bordered on the moor. Creatures with various names, often beginning with 'b' – such as bogan, boggart, bocan, bodach, bogey beasts, bogies, bogles, bogy, bugan, buggane, bugs, bug-a-boos, boogle-boos, boggle-boos and bog bears – wandered around this territory. In Shetland, it was the trow, a small mischievous creature who – despite his name – does not appear to have much connection with the larger, slow-witted troll that lives among Scandinavian mountains – and on the internet.* For all that these were tiny, they seem to have been larger than their southern equivalents, the elves and fairies immortalised in the work of the – whisper it! – Scottish writer J. M. Barrie who was the creator of *Peter Pan*. The Scottish and Irish varieties may have their origins in the tales of Tuatha Dé Danann, said, perhaps, to be divine beings conquered long ago by the Milesians (humankind).

Found on the moorland, too, were seductive creations, who were glorified and made mysterious by titles such as 'misty maiden', 'white ladies', even enticing men towards them – somewhat unrealistically – with the 'brew of the bog', a concoction linked to the wisps of mist that sometimes rose from their surface. The nature of the landscape seemed to lend itself to these kinds of beings, both ethereal and insubstantial, impossible to define. As Karin Sanders points out in her book *Bodies in the Bog*, 'there is 'something fundamentally contradictory about bogs. They are solid and soft, firm and malleable, wet and dry; they are deep, dark and dangerous; but they are also mysterious, alluring and seductive.' It is this that makes them a place of origin for presences that seem to hover between two worlds or ways of existence.

This is reflected in the way these stories were sometimes intermingled with those about the 'second sight', a gift that seemed to be particularly inherited by those who were most spiritually minded. There were deaths and losses foretold; sometimes far out on the moor but more often on the road

* Davy Cooper shared with me his theory that the troll had been brought north by Irish missionaries many centuries ago. Like those living in more southerly latitudes, they may simply have passed on stories they had heard in their own childhoods, perhaps even providing the figures that haunted their childhoods with a new name in this more northerly latitude. Peculiarly, they too have been linked with the Picts, the lost settlers of these islands before the Viking invasion, symptoms once again of the guilty consciences of those who had replaced them.

that wound across it. They seemed to be particularly associated with sharp bends on our route to Stornoway.* There was one I recall at a twist in the road leading to the village of Galson; another – as I noted earlier – beside the local hall. Both seemed to me – even at a young age – examples of the chief engineer of roads working in collusion with the uncanny and supernatural. Roads that are often empty also seem to have this surreal quality, as Samuel Beckett noted in his lesser-known novel *Mercier and Camier.*

> A road still carriageable climbs over the high moorland. It cuts across vast turfbogs, a thousand feet above sea-level, two thousand if you prefer. It leads to nothing any more … None ever pass this way but beauty-spot hogs and fanatical trampers. Under its heather mask the quag allures, with an allurement not all mortals can resist. Then it swallows them up or the mist comes down.

Often it was animals that were the bearers of bad tidings. Throughout the north of Scotland there was the water-horse or *each-uisge* stirring in any open stretches of water, but particularly moorland lochs. Known under various guises, from the *nuggle* in Shetland to the *cabbyl-ushtey* in the Isle of Man and the *ceffyl dwr* in Wales, the water-horse may have originated in human sacrifices to gods associated with water, but later became linked to stern parental warnings to

* And yet I am being unduly cynical here. One of the most odd and inexplicable experiences of my own childhood involves second sight. I recall sitting at the kitchen table and my father announcing that he had to go to a neighbour's house because there was something wrong with the man. He returned a short time later, having met the individual's wife and been assured there was nothing the matter there. Her husband was working at a peat stack in the village, helping them taking home their fuel. However, some 15 minutes after this, the old man took a turn and died.

keep children away from dangerous shorelines or deeps. Even the domestic horse possessed its own mythology. In our district, it was believed that if people saw a grey steed in their dreams, death lurked not far away.

For instance, in our village, it was believed that if a deer came to someone's gate, it was a sign that there was a loss about to come to the household. (This was at a time when there were hardly any deer in the district. Nowadays, the moor is overrun by them.) This is a superstition that Flann O'Brien – or Brian O'Nolan – satirises in his Gaelic novel, *An Béal Bocht* ('The Poor Mouth'). When the main character Bonaparte O'Coonassa is born, there are dark warnings of the misfortunes that will come to the household: 'Martin, isn't it the bad sign that the ducks are in the nettles? Horror and misfortune will come on the world tonight.'

The prospect of the end of the world is presaged later with even more bizarre behaviour on the part of the animals around Corkadoragha, a place where he is 'raised in the ashes' of a peat fire: 'I heard a cow screeching in the field with a pig's voice, a blackbird bellowing and a bull whistling.'

For all that the Irish satirist is clearly out for entirely different effects, there is something in these words that reminds us of Shakespeare's *Macbeth* with the brinded cat, the hedge-pig whining, the raven croaking itself hoarse, the horses turning wild in nature ('*'Tis said they eat each other*'). They act, too, as a reminder that the Scottish play was partly inspired by the Bard's eagerness to please his Scottish monarch, James VI and I, 'the wisest fool in Christendom' and the author of *Daemonologie* in 1597. This also includes notes from the North Berwick Witchcraft Trials in its covers. Again, there is feline activity in the reporting of this. One of Shakespeare's prototypes apparently:

> confessed that at the time when his Majesty was in Denmark, she being accompanied with the parties before specially named, took a Cat and christened it, and afterward bound

> to each part of that Cat, the cheefest parts of a dead man, and several joints of his body, and that in the night following the said Cat was conveyed into the midst of the sea by all these witches sailing in their riddles or Cues as aforesaid, and so left the said Cat right before the Town of Leith in Scotland: this done, there did arise such a tempest in the Sea, as a greater has not been seen: which tempest was the cause of the perishing of a Boat or vessel coming over from the town of Brunt Island to the town of Leith.

It is this supernatural note, frequently combining magic and moorland, that I discovered time and time again in the literature I encountered in later life. The Scottish novel *Sunset Song* by Grassic Gibbon, which I first met in my last years of secondary school, contained a few scenes that, for all that they might have been embellished with a little more learning than those I discovered in my home, would not have been out of place there. When the Guthrie family move home from Echt in Aberdeenshire to Kinraddie in the Mearns, the central character Chris comes across a wild 'half-bearded, half-naked man' on the moors, who cries out not in Gaelic but Greek: 'The ships of Pythea! The ships of Pythea!'

His words are a reminder of the Greek navigator who is said to have circumnavigated the British Isles in the fourth century. There is also the scene when, on his return from World War One, the Socialist Chae Strachan meets a soldier from a Caledonian chieftain called Calgacus from around AD 84, when these forces were battling with the Romans. Again this glimpse of the supernatural occurs on the moorland – for all that the appearance of this ghostly warrior may, as the narrator admits, be influenced by Glenlivet whisky. This is a phenomenon that also occurs in Hugh MacDiarmid's *A Drunk Man Looks at the Thistle*, where the poet mingles moorland, whisky and visions to hallucinatory effect. This potent mixture appears again and again in the Scottish literature of the early nineteenth century – from Sir

Walter Scott to James Hogg's *Confessions of a Justified Sinner*. The last is possibly the most elusive work of all. Its perspective shifts continually. Sometimes events are narrated by an 'editor' who attempts to provide an account of the 'facts' as he believes them to be. At other times, the voice of the 'justified sinner' is heard. It operates within what has been described as a 'pseudo-Christian world' with angels and demons in abundance. Spectres of light – like the Irish idea of the will o' the wisp – flicker on the moorland.* At its end, a letter (written by Hogg himself) tells of a body being discovered by peat cutters in a peat bank.

Yet it seems to me that this supernatural note is strongest in the work of the Highland novelist Neil M. Gunn. In *The Silver Darlings*, there are continual parallels being drawn between the main character, Finn, and figures from Irish and Celtic mythology, such as Finn MacCool. He is given special powers, able to 'whistle up the wind', foretell the future. He is even able to restore his mother's life in one of his first heroic tasks, walking across the moorland from the fishing village of Dunbeath to the thriving town (or herringopolis) of Wick in Scotland's most north-easterly corner.

In his novel *The Green Isle of The Great Deep*, however, Gunn does something even stranger. He links the moorland that his characters Young Art and Old Hector explore with the terrors of his own time, the Nazi regime that dominated so much of Europe during the decade in which he created a great deal of his finest work.

In this, he is, perhaps, the most prescient and prophetic of all.

* Two modern Scottish novels, *The Testament of Gideon Mack* by James Robertson and *The Bad Sister* by Emma Tenant, owe it a tremendous debt.

PART THREE

Fàd a' Ghàrraidh – The Wall Peat

CUTTING PEATS

When he went to moor to cut peat,
he'd turn over quarrels with his brother.
A feud about a forgotten football match.
Squabbles over crofting borders.
And not far from the surface,
just below a crust of turf
he scraped off with an edge of spade –
a dispute over mother's love.
And he'd dig even deeper,
slice through root and branch of heather,
recalling bruised skin and emotions,
insults they spat at one another,
till preserved there within the peat,
like the dead trunk of a rotten tree,
he'd find the source and root of it;
a deep, primeval memory
that would keep him warm in winter
for trapped within the layers of peat
was the moment that still blazed for him,
restoring all his ancient hate and heat.

CHAPTER EIGHT

Móinín Pollach (Irish) – Small Pitted Bogs

There is one turf cutter *par excellence* in Dutch history.

His name is Adriaen van Bergen, a Dutch sailor from the town of Leur. Way back in 1590, he was one of those who came up with a cunning plan to win back the fortified city of Breda in the south of the Netherlands. It involved using the peat barge which he sailed in and out of the city, delivering winter fuel to the troops garrisoned there. This was never checked by the guards, who only glanced at the ship as it made its slow way through the entry known as the Spanish Gate. Together with the nobleman Charles de Héraugièr, he suggested that this oversight might prove useful; his peat barge could be used as a Trojan horse to bring the Dutch troops into the city.

And so it happened. Acting under the orders of Maurice of Orange, 68 men clambered aboard the vessel, hiding themselves under the sods and turves of peat Adriaen van Bergen had crammed within the confines of the barge. (Ironically, Adriaen wasn't there. He had apparently slept late that day, reluctant, perhaps, to fight in the battle. Instead two of his nephews took his place.) A slow voyage followed. In late February, a layer of ice had formed on the Mark river, making it difficult to travel to the destination.

Eventually, however, the boat reached Breda. In the moonlight, men scrambled out of the depths of the peat, sending a signal to those in Maurice's force, who began their march towards the city walls. Shortly after that, the rest of the troops broke cover, casting aside their canopy of peat and emerging on the river's banks. Half of the men launched an assault on the guard; others overwhelmed the city's arsenal. Within a short time, Breda was under their control; the peat they had brought covertly through the Spanish Gate now being burned in celebration, the flags and regalia of those who had opposed them now aflame on top of these hastily made bonfires.

It is not only nations that have figures of myth. For Elleke Bal, a freelance journalist, and the rest of her family from Odoorn in the Dutch province of Drenthe, not far from the border with Germany, one came in the form of her great-grandfather, Jan Bal, a man who possessed the same name as his grandson, her father. Born in 1903, he came from the village of Erica, not far from the town of Emmen in the region. Its place name, drawn from the Latin name for heather, tells us much about the realities of Jan Bal's life. He lived in a community which had only started to form in 1863, brought together by a shortage of fuel in the Netherlands of that period. This tall, thin man was one of an army who came there; their energy and strength useful to cut out peat for households and industry, or to dig canals to bring that fuel to other parts of the Netherlands or even beyond its borders. Like those who laboured in the centre of Ireland or the Highlands and Islands of Scotland, he heaved and cut, bringing warmth and fire from the black depths of that low and level landscape, transforming it too into ground where buckwheat – often fertilised by peat ash – and other crops might be grown.

It was this that gave this part of Drenthe its own distinctive identity, one based on the peat cutting and small farms that existed there at one time, though this way of living has now given way to urban life with the horticultural industry and a major zoo together providing the foundation for the prosperity of the modern city of Emmen. There is little doubt that Elleke's ancestor, Jan Bal, would be out of place there, despite the fact there is still room for the cattle, pigs and vegetable garden he used to keep, an opportunity for him, too, to clamber on his bicycle and pedal to the village of Weiteveen ('White Bog') to work every day. As far as his family can recall, he was employed in the Bargerveen for five and a half days a week from 5 a.m. to 8 p.m. from the end of March or beginning of April to Midsummer's Day, 21 June, throughout his working life, starting when he left primary school and continuing until retirement. As Elleke tells me, year upon year the pattern was the same, its rhythms and hours as predictable as his work upon these massive stretches of peat.

'According to my grandmother,' she said, 'he worked all summer and then winter he had to recover. In the winter months he sat in his chair, smoking a pipe brimming over with shag tobacco and looking out of the window, barely speaking a word. It was as if he was recovering from a whole summer of working, trying to ensure he stayed safe and strong for the following year. He gained a lot of weight in winter, and then lost all these pounds again in summer when the work started, a constant seesaw through the seasons. At least, though, he avoided a lot of the health issues that were linked to his labours. He even lived until his mid-eighties. For many, there were constant back problems, being bowed down by the weight of the burdens they were forced to carry.'*

It was also, as Elleke told me, a life marked by poverty. During the winter months, the government supported the peat workers because they couldn't do any other work. However, despite the necessity of the work they were employed in doing, this was not a great deal of money, a pittance to protect them from hunger. Their circumstances could be seen in the way Jan Bal's home – one in which he raised most of the family's six children – only possessed a clay floor. 'It was only when my grandfather was a bit older

* In this, Jan was probably not that exceptional. In 1890, 69-year-old Pieter van Boer was encountered working on the peat bog, a place where he was first employed at the age of nine. It was noted that he was still 'doing his work with youthful pleasure'.

that they moved to a more modern stone house.' There were also the clues provided by their clothing. Elleke's other grandmother, whose father was a milkman and possessed a little more money in his pocket, recalled how she could always tell the children that came from peat families: 'They were very poorly dressed.'

And the nature of their lives determined the level of their ambitions. For all that they were intelligent children, education was not a possibility for them. It is only in more recent generations that, like Elleke, there has been any possibility of other paths for the people there to follow. She grew up in Odoorn, some 12 kilometres from Emmen, where she cycled to school every day. Unlike some of the communities in the Netherlands, such as Jorwerd in Friesland in Geert Mak's book *An Island in Time: the Biography of a Village*, it is not built on a *terp* or raised area of peat. Instead, it has its foundations in the *hondsrug*, a natural sand plateau higher than the ground that surrounds the village. It lies not far, however, from tiny Odoornerveen where the Oranjekanaal was built, its waterway linking the peat villages to the Bargerveen west of the city of Emmen. It is a location which carries its own weight of meaning for those who come from the likes of Amsterdam, where city-people have similar attitudes to those in urban Scotland. While studying journalism at university there, Elleke was continually greeted with remarks about her home place. Some would assert that the people from Drenthe are uneducated and backward. Others would ask: 'Are you from Drenthe? But you don't have an accent!'

Most of all, however, they would assert that Drenthe was still 'the countryside', 'a province of peat and poor people'. In this, they are dependent on a view of that area which held true until at least the beginning of the nineteenth century, when three-quarters of Drenthe was still regarded as little more than a wilderness, one that produced wool for the Dutch blanket industry, horses for the town and – like the Scottish Highlands and Islands – cattle that would be slaughtered in urban areas, driven to markets in places as far away as Flanders. In this, it had much in common with other areas in the Netherlands at that time. The location called the Peel, for instance, to the south, was founded by the owner of a peat company, its existence almost solely predicated on the bog that stretched across 400 square kilometres there. Friesland possessed extensive fens and marshes. Much of Gelderland and Brabant was inaccessible heathland and wilderness.

Drenthe has moved on considerably from these days, but it is still, Elleke admits, a place that has high unemployment, where she would find it impossible to follow her profession, writing on science and education. Instead, she had settled in Rijswijk near the Hague in the west of the country, living and working there. When she returns, it is to an area where there is little to remind people there was once a major peat industry. There are the place names, many of which contain words like 'moer' and 'veen' (fen), clues to the fact that the landscape they walk upon has not always been the rich, green domain it might appear to be today. Sometimes residents and visitors might come across a canal where a peat barge was once moored, taking fuel elsewhere. These are other small reminders of how peat has always been of great importance to the people of this country. For centuries, many of the people of this area lived in *terps* like Jorwerd, mounds of peatland where their homesteads might be safe from the flooding that sometimes affects the Dutch countryside. In the beginning, these were turf huts, rough and basic buildings where families lived and stayed. As time went on, they became more sophisticated: their roofs made of reed and straw, a hole (sometimes) that served as a chimney, a byre for cattle as part of the building. The area of the *terp* also sometimes grew, over the years becoming locations for churches and whole villages, 'islands', as one person described Twente, another area not far from Drenthe, 'in a sea of desolation'. Often when it rained, this landscape would become saturated, forcing people to move around its expanses by employing a barge rather than a cart, the latter's wheels becoming fixed and immovable in the mud and swampland. In its isolation, the *terp* stands in a similar terrain to the one which Antony, the guide at Clonmacnoise in the middle of Ireland, described when he spoke of his home in Bloomhill nearby: how it looked as if the tide of peatland had been receding over the generations, moving away from the slight rise in the landscape where his house stood. There were times when the entire countryside shone like glass, both the clouds and light from the sky reflecting from the water that had settled on the land.

Way back in 50 BC, a Roman soldier, Pliny the Elder – Gaius Plinius Secundus – had already noted this. Based in the northern Netherlands for a while, he wrote of how the *terp* dwellers existed:

> When the water covers the surrounding land, they look just like seafarers on a ship, but when the water recedes they seem more like castaways, the way they go hunting about their hovels for the fish that retreat with the sea.

He continues with this description, showing sympathy for them and their lives.

> They pick up mud with their hands, dry it, more in the wind than in the sun, and they use this earth as fuel to heat their food and themselves, half-frozen as they are by the northern cold. They only have rainwater to drink, which they keep in pits near the entrance to their doorway. And these peoples say that being conquered by Rome is tantamount to slavery.

Yet partly because coal was rare within the country, only mined in the south-eastern province of Limburg until 1975 when the industry finally came to an end, this landscape was of real importance to the Dutch economy. Despite the thousands of posters and postcards that display windmills, it is the case that in the country's Golden Age, roughly around the seventeenth century, they produced less than 4 per cent of the energy engendered by peat. Before the coming of imported oil and coal, it was this substance that was the black gold which fuelled Dutch industry. It was what helped builders fire up their red bricks and bakers bake their bread. It was found, too, within the limekilns within the country's borders, providing heat for that particular alchemy to occur for use in either agriculture or building.

It was even used at one point to create salt for herring, the dish prized by the Dutch. Much of the salt used to preserve and season herring was taken from the coastal land outside the dykes that protected much of the Netherlands from flooding. The peat there was clearly soaked with seawater. After they dried the fuel, in the traditional way by the heat of sunshine and the strength of wind, they burned it, mixing its ashes – or 'potash', containing 'potassium' – with seawater. This mixture was boiled until it evaporated, leaving salt grains in the pan. This method was used near the shoreline until the sixteenth century when it was abandoned because of the damage it was doing to the land and the effect it was also having on the dykes built along the coastline, leaving them more vulnerable to the lash and fury of the sea. The practice was recorded, however, in an engraving by J. C. Philipz, which was in itself based on a drawing made by Cornelis Pronk in 1745. This had its genesis in a painting that hung in the Hospital in Zierikzee in the province of Zeeland, which is said to portray people undertaking the process in Flanders during the early or mid 1550s. Clearly there is a problem in stating whether this illustration is authentic or not, as it was created over 200 years after the practice had died out and this kind of activity on the edge of the Dutch nation had been outlawed.

Nevertheless, the painting illustrates the many steps that would have been involved in extracting salt from peat, one that was copied in other parts of the world, including much of the east coast of the British Isles.

Yet through all this, the other traditional form of peat cutting went on. Industry continued to rely on it. This was especially true of the pottery for which the Netherlands is still famous. The small factories in the west of the country that first produced this in the fifteenth century were situated in areas where a plentiful supply of peat or – sometimes – wood was available to fire the kilns necessary for the trade. The industry developed later in the nation's north when Protestant craftsmen from Antwerp travelled in that direction to obtain greater religious freedom and economic opportunities than they had elsewhere, settling in cities like Middelburg, Dordrecht, Haarlem, Delft, Leiden and Amsterdam. It was there they developed the more elaborate designs that characterise their ware. These might be hand-painted with illustrations that, alongside the traditional windmills, would even have featured peat barges taking their fuel across a wintry Dutch landscape.

More controversially, peat was also used in brewing and distilling. Though it was a valuable industry in terms of the tax it accrued for the authorities, the distilling was especially disapproved of by the more sober-minded and austere among Dutch citizens. The peat-hauliers' guild in 1637 went so far as to threaten to punish anyone who was found delivering fuel to a tavern or a distillery where spirits were made, with a fine equivalent to six weeks' wages, but there is no sign that such a sentence was ever carried out. (As someone from a Calvinistic background myself, I am all too aware of the capacity of those from that background to wink and blink at sin when it suits them.) If their attitude to that form of alcohol was ambiguous, this was even more true in their view of beer and brewing. Before coffee became inexpensive enough for ordinary people to consume with any great regularity, beer, often flavoured with nutmeg and sugar, was the customary morning drink. Farmers would set off for a day of work with their crops and cattle with the best part of a flagon of beer poured down their throats. No deal between merchant and tradesman was complete without strong drink to seal the bargain. It was a habit that many church pastors complained about, including one quoted by the Dutch writer R. B. Evenhuis in his book *Ook dat was Amsterdam*: 'Men drink at the slightest excuse, at the sound of a bell or the turning of a mill. The Devil himself has turned brewer.'

From the admittedly biased viewpoint of the church pastors, there was great evidence that the peat cutters had allied themselves with Satan in this enterprise. Those in the Calvinist tradition had always believed that it was a religious duty of those who lived in the Netherlands to reclaim land from the sea, redeeming both its domain and themselves from the power and control of the ocean. It was a view that was summed up in the words of the sixteenth-century hydraulic engineer Andries Vierlingh, who declared: 'The making of new land belongs to God alone, for He gives to some people the wit and strength to do it.'

However, both the morality and practicality of this entire process were being undermined in every sense by the peat cutters, who, in their efforts to provide more fuel for the breweries and distilleries, were lowering the level of the land itself and making it more prone to the sway and encroach of the oceans. Their efforts sank the landscape. Over the entire peat-cutting areas of the Netherlands, such as Friesland and Groningen, the ground dropped by 2 metres. Sometimes, just like the centre of Ireland where the industry has prospered, the land shrank and dwindled by even more. From their position in the pulpits of the Netherlands, the Calvinist preachers could see in which direction the tides were running and made threats and warnings over what they thought might happen. The endless appetite for beer and spirits displayed by some on the edge of their congregation was performing the same Satanic work as the salt-makers had once done. It ushered in the seas themselves, sinking the land until it would be more easily whipped away by wave and storm, breaking through their carefully constructed dykes and polders, causing death and destruction to enter and destroy the barriers that had stood for centuries, protecting their nation from being engulfed. It would be God's revenge on those who had quaffed and tippled. Unfortunately, it would probably also take too many of the just and righteous along with them in its clutch and sway.

There were ample bad news and omens on which they could draw for their sermons. As I travelled around the Netherlands with the Dutch historian Karel Leenders, it was easy for me to see evidence of places where settlements had been. Karel pointed them out as we travelled. A seam of black earth. A blocked drain. A ridge in an onion field that marked where the buildings of a village called Polre had formerly stood, built in 1250, drowned in 1583. A line of sand dunes guarding the seashore concealed evidence of the existence of a town called Reimerswaal that disappeared near the end of the sixteenth century. Its existence had come to an end as the result of two storms: one in 1530, another some decades later when water flooded the area. The settlement's extinction had much to do with the peat cutting that had gone on for some time behind the town. Moorland was no longer in place to soak up the rainfall that so frequently showered down on the landscape. Instead, much of the 'brown gold' had disappeared, its wealth and depth used to produce ceramic plates, a flagon of ale, a bottle (or more) of gin.

There was a moral failing in all of this – a failure to safeguard the world which their forefathers had provided for them, an inability to defend and protect an inheritance given by the mercy and grace of God. There were ample opportunities, too, for preachers to write sermons lecturing the faithful on their weaknesses and follies, like the great floods that took place over the centuries, at regular intervals such as 1731, 1798, 1808–09. Each of these disasters gave further evidence that the Almighty was discontented with the people of the Netherlands, especially when one considered that some of its prime causes were eminently preventable, especially in a preacher's view. The drunkenness of those who swallowed gin and allowed beer down their throat was doing great damage not only to the country's moral standing but also to its landscape.

When one comes across a Dutch peat bank, it is easy to see there might be some basis for their complaint. These are not the narrow ledges of land on which my father and I stood when I was young, carefully taking out peat for our household fires. Nor is this the short concentration of weeks when I would go out after school or on a Saturday, lifting peat or gathering it. Instead, this was the heavy, harsh work the likes of Elleke's great-grandfather faced throughout a considerable portion of the year, rising at twilight and returning home in a darkness as full and complete as the lower layers of the peat he was paid to cut.

Ironically, it was in a family theme park in Drenthe – one which proclaims 'Step into the world of peat and explore the all-new Veenpark' – that I encountered what Jan Bal must have experienced. Within its borders, there was a children's playground with swings and a wall they could clamber, an interactive display where they could learn about the environment that had at one time stretched across this region. A white building in the shape of a harmonium also celebrated that instrument. I must confess I thought at first it was an accordion, its outline recalling to me the occasion I had been out gathering peats on the moor and a local musician had decided to take one out from its case to entertain us. (It might have been a Strip-the-Willow that he played. It certainly encouraged me to strip the peat banks more quickly than I might otherwise have done.) There was, too, a train which puffed its way past a community that represented the history of homes and other buildings in this region – from a traditional turf hut to a church, a school to a rather expensive restaurant.

With my friends Roel and Aleid Bosch, I stepped across this damp and wet terrain warily, coming at last to a bank where peat was still cut. It was there we met an elderly, energetic man by the name of Rolf van der Vaal. It was his task to show visitors how peats had been cut on that landscape for centuries. A stocky, red-faced individual with a ready smile and white moustache, he did this eagerly, speaking of how six men would normally be employed on this wide path, about the breadth of a household table and a staggering 6.5 metres in height, across this edge of moorland, working – just as Jan had – long hours throughout the peat-cutting season.

'They could not do it any later than that. They need three months without frost after cutting for the peat to dry. Frost is the kind of weather that causes peat to become hard and brittle, removing its capacity to give off heat. Frost sometimes arrives here from late September, early October; that's why the season is so short.'

Yet they would be extremely active in that time, employing – what was to my eyes at least – a bewildering array of tools, including an instrument shaped like a cricket bat that cut the peat, levering it from the earth in long, thin slices. Between 1866 and 1966, when they finally closed the peat-cutting company within the boundaries of the modern-day Veenpark, the workers cut 1,350 hectares of fuel, harvesting 10,000 tons every week of the season – a total they were compelled to reach. It was a task that involved whole families: wives and children scurrying back and forth, laying down little tepees of peat that resembled those I had created at home. For this, they were paid a small amount of money, some 75 cents for 1,000 turves or 100 piles. Their efforts were often watched over by an inspector employed by the consortium of farmers who, together with an investment company, owned the land and supplied any mechanical equipment that might be required. This overseer would live nearby, staying at a bed-and-breakfast.

These were the working patterns of this area until the early years of the twentieth century. They are illustrated even by the great Dutch master Vincent van Gogh who, during his time in Nieuw Amsterdam in Drenthe in and around 1883, spent a considerable amount of time sketching farm workers, particularly the women who were gathering peat on the moor, as my own female relatives sometimes did, with a wheelbarrow by their side. He returned to observe this landscape time and time again, finishing a painting entitled *Two Women in the Moor*, telling his brother Theo, too, of how: 'Yesterday I drew decaying oak roots, so-called bog trunks (being oak trees that have been buried under the peat for perhaps a century, over which new peat has formed – when the peat is dug out these bog trunks come to light.)'

He also painted a peat boat while in the area, a man wheeling his barrow down a canal bank, a woman hunched on the barge, and a couple of landscapes that would have been much appreciated by the island crofters of my youth, one showing a peat bank, another consisting of a large peat stack and two buildings.

A few years later, many of the peat banks that Jan Bal and his friends worked at had become mechanised. This happened after the revival of the peat industry during the First World War, a phenomenon which repeated itself in the Second World War when, once again, the Netherlands could not rely on largely imported coal. In 1920, the overseers introduced a variant of the conveyer belt to the industry. Green, gleaming and bearing the name – S. de Graaf – of the firm that had owned these peat fields in its latter days, it still remained next to where men were now employed within the theme park to demonstrate the traditions of harvesting the fuel. Rolf drummed its side with his fingers, proud that – just before the centenary of its creation – it was still able to work.

'A high-quality English improvement,' he declared. 'Best engineering in existence.'

Beaming, he raised a cynical eyebrow, as if to suggest that this was no longer the case. He showed, too, how it worked, the manner in which it first compressed the peat, oozing water, into two briquettes that travelled the length of its black belt, making it easier to dry as well as performing the task that wives and children had done when they worked not far from here in van Gogh's time of staying in Drenthe. It did this much more quickly than human fingers could ever manage.

'Forty thousand tons a day…'

The peat would be stored in a nearby shed before being taken on a barge to the breweries and households where it was later burned. Looking at its dark shadow within that building made me think of how the likes of Jan Bal must have worked in the peat fields of the Netherlands. This was not the occasional seasonal work of my ancestors, who could break the tedium of their moorland labours by taking time off to work with sheep or the crops that grew on their croftland, spreading dung and seaweed on its acres, ploughing the earth in dark, straight furrows which they could witness transformed into green a few months later, the small miracles of that change taking place. Instead, this was the back-breaking boredom of going out to the same moorland day after day, working for an employer who would provide them with a pittance in return for both their hours and their sweat, the poverty of body, mind and soul.

There were occasions when this was interrupted by anger. This was particularly true in Friesland, on the north-western edges of the Netherlands. It contains communities where the place names continually remind the visitor of the nature of the earth that once lay below the tarmac and pavements. One bears the name Heerenveen (the Lord's peat). Its football team seems to be kitted out in the kind of special costume that should only, perhaps, be worn on Valentine's Day. Red 'love-hearts' adorn its strip, as if some moonstruck teenager has been allowed loose on the design, bearing blue – for the occasional stripe – and red felt-tip pens in their fingers. Both the *Guardian* and international sporting magazines have often in the past regarded it as among the worst footballing strips in Europe, akin, perhaps, to my native Scotland's pink outfit or some of the more dramatic splurges of colour that were seen on men's torsos when they played soccer in the seventies.

Yet apparently it is nothing like that. Instead, these shapes are supposed to resemble the lily-pads that I recall from childhood, stretching along the edge of moorland lochs, perched above the water as if they were waiting for frogs and toads to leap and dance upon them, turning a shade of rose when autumn came to these empty acres – for all that they are not as bright and garish as those outlined on the jerseys of the players of Heerenveen, nor the same shade of crimson as those found on the souvenirs and scarves of the team's supporters.

There was a time when other items of that colour flapped and fluttered within Friesland, especially at its most south-easterly edge in communities like Nij Beets and Tijnje in Opsterland – the red flags and towels that were taken out when a strike started in the peatfields on which their communities were founded. This was an event that occurred many times in the nineteenth century, with the industry coming to a halt on many an occasion between 1810 and 1925, when the last stoppage occurred. It was a workforce that was truly international in its origins, comprising migrant workers from Lower Saxony in Germany as well as Dutch provinces like Holland and Friesland, yet they came together on many occasions over the years, particularly from 1856 to

1888, when, led by Socialist agitators, the periods in which they were away from their peat works lengthened. In the beginning, they only stayed away for five days. By 1869, the periods had lengthened until, through more organisation, times out on strike lasted between 19 and 27 days. In 1890, they fought for and won a commission of enquiry to examine the conditions of working people in the area. In 1894, their trade union was recognised.

In the course of these disputes, they created a radical tradition that – for a time – was of great importance in the Friesland area. Its legacy could be seen in the fact that 25,000 men took part in strikes in the region from 1925 to 1932. The peat-working communities were also responsible for the election of the first Socialist representative in the Dutch Parliament, Ferdinand Domela Nieuwenhuis, the man to whom the peat museum in Heerenveen is dedicated. Born in Amsterdam, he was an Evangelical Lutheran preacher who lost his faith after working in various towns throughout the Netherlands. In 1881, he was the spokesperson for the Social Democratic League (*Sociaal-Democratische Bond*), inspiring and helping groups like the peat workers to go on strike. Within Parliament, he protested about the suffering they endured, pointing out that:

> peat-cutting required a lot of strength, and not only men are doing it. Even women – some of whom gave birth less than 14 days ago with their skirts tucked into their pants, stand there in the burning sun with their backs bent, with an iron shovel, with their young children in a small cage, so they can keep an eye on them, or while other children – either or not their own offspring, are watching them. Even boys 12 years or younger are being put to work in these circumstances.

There is little doubt that even outside the context of his role as a spokesman for those employed in the peat industry,

Nieuwenhuis was a firebrand. In 1887, he was imprisoned for a year after he – allegedly – wrote an article insulting the Dutch royal family. Even after he was elected to Parliament, he became increasingly drawn to anarchist ideas, and only remained there for three years. In 1896, largely because of his pacifist beliefs, he was expelled from the Second International Congress when it took place in London.

With his full white beard, in the statue that stands to this day in Amsterdam, he is still seen by many in Dutch society as a figure of myth and heroism, much as Elleke Bal views the life of her peat-cutting great-grandfather Jan. As a peatland working-class hero, there appears to be no one quite like him in the rest of Europe. As much philosopher as politician, he acted as spokesperson for the peat communities to be found within his nation's boundaries, shaking his fist at those taking advantage of their sweat and labour, drawing attention to the cold and injustice they suffered at their work.

His physical similarity to that other grand old man of the left, Karl Marx, is striking.

So, too, for anyone who studies them, is the bright and blazing anger of his words.

CHAPTER NINE

Bluster (Shetland) – Rough Peat

There was only one occasion in the year when my dad turned up at my primary school.

These were not parent–teacher nights. Neither were they times when he was summoned to the classroom by the headmaster, struggling to deal with my misbehaviour. Instead, a cloth cap perched on his head, a smart tweed jacket on his back, he would arrive in the school playground, having forsaken his loom and croft for an hour or two, to pay his dues to the Galson Estate, the landlord who owned much of the north end of the island, moors included. He'd take a few crisp notes from his wallet and pay the year's levy to the factor, a Stornoway solicitor, Douglas Kesting, a man who looked and behaved not unlike himself. Both were quietly spoken, thin, sallow-skinned men with dark hair combed back from their forehead in the style of the time. They acknowledged each other with a nod and the passing of a receipt before the next cloth-capped, tweed-jacketed man took my father's place in the queue.

And that was it. After that, Galson Estate retreated from all visibility in our lives. It had not – I was dimly aware – always been like that. Before the First World War, the estate had been all too visible. There was even the church elder William Murray to remind me of their presence. He had once been the child who had had a gun fired at him for having a rabbit in his possession. There had been other indignities, some of which my father had informed me of when we sat together beside the fire on these all too frequent nights when storms rattled the windows and electricity failed. But that was then, and this was now. From the pages of the *Stornoway Gazette*, it was easy to tell that the tenants of the Galson Estate were fortunate. Relationships elsewhere on the islands between estate and residents were not always so distant and detached. Further afield, there were villages and districts where there were frequent clashes between locals and those who worked for the estate. This was especially the case when the latter decided – in their wisdom – to employ a clutch of former soldiers to police their domain. It led to ripped nets and – even worse – torn relationships with the people of the community, tensions escalating as one senseless reprisal followed another, becoming headline stories in both the *Stornoway Gazette* and elsewhere. Sometimes these involved scenes that would normally be reserved for James Bond films, with high-speed boat chases, landowners resembling Ernst Stavro Blofeld and 'heavies' swaggering around with black leather jackets and iron studs. These tended to be, of course, the estates in which rivers streamed full of the silver of salmon, or where the spiked antlers of the occasional stag could be seen. There was a failure of the squirearchy of the estate to realise that trouble might be cured by a word from the mouth of one of the locals they employed, a simple question like the following: 'You speak the Gaelic. Could you go out and talk to these lads?'*

Terrified of blood spilling, even if he suffered a cut finger, my uncle Norman would have been the ideal candidate for this task. He worked on one of the estates for a few years. Occasionally he would speak about this, helping visitors from the south to catch salmon by rowing them out in one

* Hector Macdonald in *A View from North Lochs* had his own original take on the island landlord. After outlining the natural enemies of the crofter, which include the sea, stag and corncrake (for disturbing the poor man's sleep), he goes on to declare that 'the most voracious predator of all' is 'the landlord... They are at their most dangerous in August when they migrate north in loud noisy flocks. Their brief sojourn in the cold regions often proves costly for the peasant, because the landlord is obsessed with the peasant's basic food – salmon.'

of the estate's boats or taking them to a good spot on the riverbank. Those in charge of the nation's industries might be mentioned, a double-barrelled name or two, all members of the syndicate who owned the place. It was the Spartan conditions in which they lived during their stay in the Isle of Lewis that most impressed Norman. 'They weren't exactly luxurious,' he would say. 'A wee bit like going back to boarding school, one fellow told me.'

This was the impression, too, of David Profumo, who stayed in the same lodge sometime in the eighties. Son of the former Cabinet minister, he travelled to the Isle of Harris for the first time in the mid-seventies in the company of his father-in-law's family, the Frasers. It was a visit that inspired David to write the novel *Sea Music*, which is about the arrival in the fifties of a young public-school boy, James Benson, to a remote Gaelic-speaking island in the west. A devastating portrayal of the men who accompanied him on this journey, his businessman father and his right-wing Conservative friends, he also shows the people of Harris in an extremely kind light. The lodge-underkeeper Alec Campbell and especially his aunt Rachel are portrayed as people who are both at ease in this environment and kind to him. This is in contrast to James's father and his cohorts who see the island as essentially a playground in which they can indulge their huntin'-shootin'-fishin' pursuits, treating both the natives and the land in which they live with contempt.

Some of this chimes with David's own perception of life in the islands. Talking to me from his London office, he spoke freely of how the Isle of Harris with its coastline and moorland was a place of 'emotional refreshment' for him, one in which he could 're-energise' himself before he returned to the capital or even the 60 or so acres of rural Perthshire he owns. Regarding it as 'one of the last wilderness areas' in Europe, he enthused about its sense of space, the feeling of community, Gaelic tongue and the strength of the culture that existed there, 'almost timeless in a way'. As an individual, he was glad of the way that the background of his in-laws had allowed him entry into the homes of the villages there, especially Rodel at the island's most southerly point.

'There's something about a doctor that allows him entry into that world. They regard people who do that kind of work with both warmth and respect.'

For all his enthusiasm, David acknowledged that not everyone appreciated the landscape of Harris. 'It's quite Janus-faced,' he said, 'in a couple of ways. On the east side, there are the rocks. More stone than moorland. On the west side, there is the sand. All in proximity to one another. And the weather can be quite Janus-faced too. One minute it's wet and stormy, all bitter wind and damp. The next dry and calm. I can appreciate why people either take to it or not.'

In David's case, there was – from the beginning – little doubt about his reaction. His existence took on a new direction during his time in Harris, leading him – among other roles – to become the fishing correspondent for *Country Life* magazine. His enthusiasm for rod and line can be seen in the way he became the editor of *The Magic Wheel*, an anthology of writing about angling, and travelled widely both throughout these islands and elsewhere to pursue his craft. However, if you were to ask him about his favourite place in the world, there is no doubt about his answer.

'Oh, that's Harris. I love its dinosaur hills, its moodiness, how the north wind blows from every direction. I love walking across the sands. And then there's its people. The elegance and gentleness I saw on the island. It's restorative.'

There is in his memoir *Bringing the House Down*, about life in the Profumo household both before the scandal that ended his father's political career and afterwards, a wonderfully moving passage that deals with the family's reaction to the death of David's mother, the actress Valerie Hobson, and how they scattered her ashes after she had been cremated. Sensitively written, while it is set in Highland

Perthshire it tells much about David's relationship with moorland, whether in Harris or elsewhere:

> Up on the moor behind the house is a big, pale stone where I often sit and watch the glen. Just before noon, on the first day of the new year, I drove my father into the field below, so he could see clearly where we were going to stand on the brae above him; I poured him a flute of Dom Perignon ('Miss Hobson only drinks Champagne'), then Mark, my stepbrother, and I went up the hill through the smirring rain, and stood either side of the rock.
>
> We had no elaborate cinerary urn — just a plain plastic container. For a brief while we stood there, waiting for the sun. It was chill and clear, but it appeared as if on cue, so that she was perfectly lit as the ashes whirled up into the grey wind, towards the peaks of scree and the snowfields above, a last, swirling signature in the air.
>
> In the field below, my father managed to raise his glass in a salute. The clouds pass, but the sky remains.

Donald Macdonald was another man whose life, like Profumo's father-in-law, managed to bridge the difference between crofthouse and the estate. He grew up in the village in the south of Harris Gaelic speakers refer to as An t-Òb (The Obbe) but is known in English – after Lord Leverhulme – as Leverburgh. He first became what he termed a hunter-keeper in the early sixties when, as a 21-year-old, he arrived at the Luss Highland Games near Loch Lomond. The Blair Drummond Estate was advertising for young men to fill this role. In an interview with the head gamekeeper, Harry Mitchell, he was asked the question: 'Apart from poaching, what experience do you have for this job?'

Donald smiles at this memory, glad he does not need to confess the many times – from the age of 13 or so – he went out to catch a salmon or bring home a deer. There were also occasions when wild duck or shag was the target; shag, known as the *sgarbh*, was regarded as a great delicacy in the islands. Instead, a hand is offered; a job offer accepted; a poacher transformed into a gamekeeper.

After that, he worked for a few years in Blair Drummond near Stirling, a place that has seen a number of transformations over time. Once it was where the Enlightenment figure Lord Kames turned the carse, the low-level area beside the River Teith, from waterlogged moss to good, agricultural land. Nowadays it is known as the site of a safari park. Lions and tigers prowl where deer and other prey were once stalked and hunted; penguins and rhea, related to the ostrich, flap where there once was grouse. Following this, he worked for the Morrison or Margadale family, keeping a weather eye on the salmon, grouse and woodcock in the Rhinns of Islay, a part of Scotland that was not unlike his native Isle of Harris, with its mix of moor and Atlantic coastline. His sense of ease in that environment may have been partly caused by the fact that the household, which included the Tory MPs Peter and Charles Morrison and Dame Mary Morrison (who was Woman of the Bedchamber to Queen Elizabeth II for 50 years) possessed Hebridean forebears. Over 250 years before, an ancestor bearing the family surname walked from South Uist in the Western Isles to Middle Wallop in Wiltshire. His reward for his endeavour was to marry his employer's daughter. Sometime later the family made its fortune by storing black crepe fabric in anticipation of the death of King William IV in 1837.

There was only one aspect of that life that bridled occasionally with Donald.

'They always referred to me by my surname. Macdonald. The children used to call me Mr Macdonald until they reached the age of 16. After that, it was just the same as the rest. Macdonald.'

He laughs at this memory, emphasising that there was little friction between the estate and the community, something that – according to a friend of mine – still persists today. ('Alasdair Morrison's lovely,' she smiled, referring to the present landlord. 'Down to earth and easy to get on with.') The only trouble he recalled was not with the locals, who – like their Harris counterparts – would sometimes poach a salmon for the pot, but with those who came occasionally to stay in the island's hotels.

'Some of them would try and take advantage, grab a lot more than their share.'

This was not the reason he moved to work for the Barvas Estate in the Isle of Lewis in 1973. Instead, there was the proximity of family, the possibility of even an occasional drive down to Harris at the weekend. He still works on the estate, once belonging to the Duckworth family* but now owned as a trust by the people living in the area, guarding the salmon in the river, keeping an eye on the deer which – over the past few years – have become increasingly common in the north end of Lewis, intruding into croftland even near the island's coast. He admits, however, that over the last few years his work has been less arduous. Poaching, for instance, is much less frequent than before.

'There's a lot of reasons for that,' he says. 'Salmon is now much more often found on people's plates these days. No

* At one time the estate belonged to John Talbot Clifton, who followed in a long tradition of eccentric Highland landlords. An inveterate traveller, he was known for shooting animals and eating them. He shot many that were hitherto undiscovered by science, including a wild Siberian sheep, now known as Clifton's bighorn, and a Canadian marmot. He also dined on the carcass of a mammoth that was found frozen in the Arctic permafrost. The Barvas Estate includes the island of Sulasgeir where the men of Ness in Lewis go to harvest the guga (gannet chick) each year. I have no idea whether or not he ever consumed this, but, given his culinary history, I consider it extremely likely.

one any longer has to go out to net their own. Even venison can be bought from the butcher's shop.'

And that is true. Even when I was a boy in Ness, there were nights when the unlocked door of our crofthouse might open and there would be scuffling noises in the darkness. Step into the kitchen and a plastic bag would be on the table. Within it, there would be a chunk of salmon, its skin silver and sparkling, its flesh a deeper shade of red than those pale specimens one sees on supermarket shelves. Once or twice it might have been a moonlight visit from some of Uncle Norman's former colleagues on the estate, providing him with a gift on their journey home. However, it was far more likely to have been an act of generosity from one of the village lads who had poached it from a river elsewhere on the island. A few days later and there might even have been a wink and a smile to confirm this, a few words that mentioned our visitors during the night.

There was certainly no condemnation of this crime. As well as helping the economy of these often marginal communities, poaching was also the source of much of their cohesion and conversation. The following morning and afternoon would be spent discussing the exploits of the night before, how they had scrambled over peat bog and bank to evade capture, how one woman had prevented her husband being caught by tucking the evening's catch underneath her skirt. In fact, poaching of this kind was regarded as hardly a breach of the law, but almost as a sacred ceremony, a natural entitlement and everyday sacrament that was extended to the people who lived near a Highland estate. There were even a number of gamekeepers who shared this attitude, ensuring – in most cases – that providing the behaviour of local crofters was not excessive in terms of what they caught, their offence would be blinked at and ignored. (I can think of one exception to this – when a family in our community positioned a net at the mouth of the Dell river, preventing the salmon from swimming upstream. This was going too far, largely, it has to be said, because of its effect on others who, a little upstream, fancied capturing and netting their own salmon.)

There is a reason for this. Many never recognised the rights of individuals, whatever their merits, to own land, its acres belonging – in a profound and meaningful sense – solely to God. (I even recall having a conversation once when someone argued that the very word 'landlord' or the title 'Lord' was a form of blasphemy, taking the name of the Almighty in vain.) It may even be this divergence of views that explains to me one of the major differences between the Scots and English attitudes to the ownership of land. It is only in newspapers down south these days that I read reports mentioning the words 'trespass' and 'territorial rights'. The Highland Scots, in particular, have little understanding of the notion. This attitude even extended to their own crofts, which were rented but never their own possessions.

Apart from religious objections, the reasons go back to the old Highland and Island clan system. The clan lands were never ground that was 'owned' by the chieftain. Instead, it was held in trust by him. In terms of the earth below their feet, whether it was moorland or fertile soil, *is treasa tuath na tighearna* – 'the people are mightier than the Lord'.* A few centuries later, this phrase became one of the slogans of the Highland Land League, a group that pointed out that this was an 'English' concept of land ownership that had been exported north rather than one that belonged to the Highlands and Islands. Several of their representatives, members of the Crofters Party, were elected to Parliament in 1885, serving constituencies like Ross and Cromarty, Argyllshire, Inverness-shire, Caithness and the Burghs of

* I am aware there are differences in attitude even in the north of Scotland, such as Orkney and Shetland where Udal law applied. Much of that law now applies to the foreshore, however, and not to the moor.

Wick. The following year, the Crofter's Act was passed. This provided security of tenure to crofters. They were no longer to be turfed out of their own homes on a landlord or factor's whim. Despite this, there were still tensions between landowners and those who lived cheek by jowl with them, possessing tiny scraps of land. This could be seen after the First World War. The people who volunteered for the conflict had been promised land of their own after the battle had been won.* Impatient for this, there were land raids in several locations. They included one of the nearest villages to the one in which I grew up. The farm at Galson on the west side of Lewis was broken up to form croftland. The descendants of some of my fellow villagers would often be found within the houses there, asking about their relatives each time you stepped within their doors.

Yet for much of my life, it was through an abundance of comic stories that the rivalry between croft and estate was expressed. Sometimes the *Stornoway Gazette* contained front-page stories with headlines like 'Salmon Poacher Fined £15. Caught After 3 Mile Race'. The report continued with the 39-year-old culprit declaring his innocence by protesting, 'I wasn't poaching. I was only running with other people on the moor.' There was another time when a few individuals were caught red-handed with a stag they had just killed. The evidence was taken away by the Stornoway police and hung in the cold store of one of the local butcher's while they prepared the rest of the evidence for the forthcoming trial. A short time later and the court case was aborted. The cold store had been broken into and the deer

* Apparently my great-grandfather had been very cynical about this promise. 'It'll be the dead who'll build their houses on it,' he declared when the minister encouraged them to sign up. He was proven to be correct when the first homes to be built on the newly acquired land were 'widows' houses', paid by the pensions given to those who had lost their husbands in the war.

removed. Only the stag's antlers – complete with a label complimenting its flavour – remained. It was a response that a former poacher from the Isle of Harris, Sammy Macleod, would have been very disappointed with. In his youth, these horns were almost as important a part of the haul as the meat. He used to collect the points from the antlers, or 'tusks' as he called them, and, hiding them in Smarties tubes, send them on to a shopkeeper in Oban, who transformed them into the handles of kilt-knives and paid him £60 to £70 a time for his efforts. It was a task made easier by the gullibility of some of his friends. He had told them he needed antler-horn to make a necklace for his girlfriend, boring a hole in each one and stringing them together.

'I wouldn't mind,' he laughed, 'but there was no young lady in my life back then. I was only 16 and innocent.'

His chuckle undermines his words, giving me a sense that 'innocence' was not a condition to which Sammy ever aspired.

It was a time that needed quick wits, as Sammy recalled. He, his father and another man had gone out one night to obtain some venison for the family croft. They were poised behind one of the island's many boulders, a stag in their sights. Just as his finger strained upon the trigger, another shot was fired. They looked out into the darkness to recognise another group was out on the moor with them, an 'official party'. They were making their way across the moor, a snipe panicking before their feet, flying in the direction of the sky.

'I've never run so fast in my whole life. You'd have thought I was training for the Olympics.'

My friend Maggie Smith recalled how, after catching a salmon at a river-mouth, her grandfather had been disturbed – not by a snipe or curlew but by the approach of the local gamekeeper. A moment later and the fish was hidden below his jacket on the ground. Maggie's grandfather's trousers were tugged down over his knee-caps as he pretended to crouch and defecate in the bracken. 'It's a sad day when a man can't even do this without the gamie coming around,' he complained. Another of her grandfathers also left the drawer below the kitchen table empty. The reason? It was a quick hiding place for plates. If they were ever eating salmon and the gamekeeper came around, they could be hurried and concealed there.

'You had to be aware of his presence all the time.'

It was only, however, during my late teenage years that the issue of land ownership in the Highlands and Islands began to be treated seriously once again. This time in the mid-seventies was when a rabble of oppressive landlords throughout the Highlands and Islands stumbled into the headlines, upsetting those of us who are members of the hoi polloi. Step forward Dr John Green, who became known as Dr No in Raasay and elsewhere for objecting to the coming of a ferry to the inner Hebridean island on the grounds that it would spoil his view. (He only travelled there once during the decade and more that he owned the house and access to the harbour.) Come into the spotlight Hereward the Wake, descendant of the eleventh-century opponent of Norman rule, later immortalised in the track 'Rebel of the Marshland' recorded – centuries later – by a 'pagan' or 'Anglo Saxon metal band' called Forefather. He would gain his own legendary status by becoming the 'Rebel of the Moorland' by objecting to a road running close to his home in Amhuinnsuidhe Castle in North Harris. Apparently, the hordes travelling back and forth on their daily commuter run to the sands of Hushinish would disturb his privacy.

And then there was, in the starring role, Lord Burton of Dochfour, landlord of the Cluanie, Glen Shiel, Glen Quoich and Dochfour Estates, member of the roads committee for Inverness County Council. Allegedly the model for the soft-edged series *Monarch of the Glen*, based on Compton Mackenzie's novels, Michael Evan Victor Baillie provided a great deal of raw material for those questioning the landlord system operating in the Highlands and Islands. He was the scourge of the many forces with which he came into contact in his daily life. His enemies ranged from salmon lures to rabbits, cats to the Kintail Mountain Rescue Team, buzzards to busloads of fishermen from Glasgow. He even took exception to cars breaking down near the gates of his home, slamming the bonnet down on one unfortunate mechanic's fingers.

There is little doubt that these characters came from a long line of unloved and unlovable eccentrics. They included the notorious Lord Brocket, the Nazi sympathiser whose land was raided by the Seven Men of Knoydart after their return from the Second World War. However, the difference between that time and the seventies and eighties can be explained in a number of ways. First, the nature of the people of the Highlands had been altered by both the coming and consequences of the 1948 Education Act. The men and women who lived near these vast estates were now educated long beyond the narrow limits of their forebears. They had a confidence and level of articulacy their predecessors (largely) lacked. They also possessed powerful supporters.

These included the 7:84 Theatre Company, so-called to remind its audience that within the borders of these islands '7 per cent of the population own 84 per cent of the wealth'. Their play *The Cheviot, the Stag, and the Black, Black Oil* not only revitalised theatre in Scotland, it also brought the politics of land ownership in Scotland to life once more, performing this miracle before both school pupils like me – present in the audience on 28 September 1973 – and adult audiences as far apart as Portnaven Hall in Portaskaig, Islay and Stromness Community Centre in Orkney. They did all this with a mixture of means that mingled pantomime with the songs normally heard in a Gaelic ceilidh, historical fact and rumbustious knockabout, absurd creations like Andy McChuckemup, an imaginary Glasgow property owner and real-life characters like Dr Green of Surrey (and Raasay!) who was:

in no hurry
for a ferry
to cross the Sound.

And then, too, there were newspapers like the *West Highland Free Press*, based in the Isle of Skye, which both caricatured and castigated the old landlord class mercilessly.* They did this particularly with Burton of Dochfour, whose thin skin was punctured time and time again by the pen of Brian Wilson, the *Free Press* editor and later Minister of Energy in a Labour government. Sometimes his name was linked with Hereward's, when, for instance, under Burton's chairmanship, the roads committee of Inverness County Council considered the small matter of whether public funds should pay £40,000 for the swirl of tarmac around the castle owner's humble abode. On 9 March 1973, its headline read: 'Not So Fast, Worthy Lord Burton'.

Burton rode to his friend's aid in supporting this enterprise. His energy summed up in the opening paragraph where the newspaper's staff reporter wrote: 'Hereward may be the Wake, but Lord Burton was not the Unready when the roads committee of Inverness County Council met ...'

Today, a scenario like this would be much more unlikely, if not impossible. Since 2003 and the Land Reform Act introduced by the Scottish Parliament some four years after its inception, many of the communities in the Highlands and Islands have purchased the land in the vicinity of their homes. They include Ness or the Galson Estate, where my father, if he were still alive, would no longer have to pay cash

* More recently, there has been the work of Andy Wightman, the land reformer. His efforts both as an MSP and writer have been considerable and admirable.

to an outside landlord on his annual visit to school. The people of South Uist, Eigg, Sleat in Skye, Assynt in Sutherland also have laid claims to the community in which they live. Among those who are in this position – to no doubt the late Lord Burton and Hereward's eternal disapproval – are those who live in North Harris. This followed a unique deal when the landlords of that time, the Bulmer family, sold the land and castle in two different and distinct lots. One, involving the castle and the salmon river that Hereward's ill-favoured road zigzags across, was sold to a family called the Scarr-Halls. The other was bought by the population of North Harris.

It is the latter that is managed and run by individuals like Gordon Cumming. He has lived on the west side of Harris for 22 years, working in the same role for the neighbouring Borve Lodge Estate for 16 of them. It is, however, a very different task from the one he experienced there. The organisation of which he is part has to perform a large number of roles. Over the last few years, it has assisted in the creation of social housing in an area where the population is in deep and sharp decline. It draws much of its income from fish farms and aquaculture yet it needs to be sensitive to the ecological concerns the industry has generated over the last few years. It is required to cater for those who are resident to the area and those who come to visit this stunning and humbling part of the world.

And both stunning and humbling it undoubtedly is. In his poetry, the Welsh preacher and writer R. S. Thomas used to describe the moorland in spiritual terms, calling it either a 'cathedral' or a 'church', which he entered:

> on soft foot,
> Breath held like a cap in hand.

There have been times in my experience when this has not been the case, when under the lash of rain and in the cloying damp that sucks at a person's feet, it is more likely to be a curse than a prayer that comes to my lips when I walk across moorland. This has rarely been the case in the north of Harris. I have stood alongside the high, perpendicular rock of Sròn Uladal (or Strone Ulladale, as it is known to many) and found it easy to convince myself – however temporarily – of the ascent of angels. I have scrambled down into the once cut-off and remote village of Rhenigidale and seen an eagle floating, its wings and feet on fire. I have scrambled to the top of the Clisham, the highest hill on Harris, and felt my eyes dampen and grow wet, astonished by the wonder of the landscape that lies below. It is a place where, to use the words of R. S. Thomas, on numerous occasions: 'What God there was made himself felt.'

And yet, in contrast to this, there are some parts of the Highlands and Islands where God and a sense of the common good appear not to be 'felt' at all.

PART FOUR

Mòine Dhubh – Caoran – Black Peat

LOOKING THROUGH A GLASS DARKLY

Most of the artistry we possessed
was inevitably in black,
matching the binding of Bibles,
the widow's dark
clothing as she bent to shovel ash,
an occasional fleece among our Blackface flocks,
blood-splattered beaks of crows,
jagged profiles of harbour rocks,
soot shrouding backs of fireplaces,
murky shadows of a moorland loch.
And we expressed this in our conversations
where we made predictions that were stark
and grim, savouring the grace of clouds,
but, above all, rejoicing in the black
of peat we stacked beside our homes
each summer, fixing turf in exact
and taxing patterns, shaping moss and bog
into designs that might distract
us from sorrows, both bric-a-brac and artefact.

CHAPTER TEN

Blue Clod (Shetland) – Dark Peat

'I'll tell me ma when I get home …'

The guard in Dunrobin Castle whistled between his teeth as he stepped the length of the drawing room, keeping a weather eye on the visitors that had arrived at this eastern edge of Sutherland in the north of Scotland one afternoon in early October. He would occasionally glance, too, out the window, envying, perhaps, those who were outside on this unseasonably warm autumn's day in the magnificent Italianate garden that stretched down towards the coastline. There would be a skip of his tartan trews, a brisk tap of his shoes, as he changed tempo, whistling an old Gaelic air to accompany his patrol down the carpet. It was clear even from the constant changes in his choice of songs – '*A rìbhinn, a bheil cuimhn' agad*' one moment, an English song the next – that he was a restless soul, kept in check by the income he received as much as his ancestors might once have been restrained by the 2nd Duke of Sutherland, whose portrait was on the wall. When I spoke to him, a muttered 'Fine day', he clicked his head in my direction and smiled – 'Aye. It is that.' – before continuing with his stride and jaunty rhythm, that Irish song restored to his lips.

'I'll tell me ma when I get home …'

I must admit I shared his restlessness. For all that I had paid my entrance fee to Dunrobin Castle, I was reluctant to obtain the full value of its cost. I dodged the tea-room, only stepped on the outer edge of the garden. I felt oppressed by its grandeur, the portraits of the often unnamed ancestors of the Stafford family who owned this fairy-tale palace, the sense it had of being the width of continents away from its immediate landscape, the towns of Brora, Golspie and Helmsdale nearby. Their foundations were not dug out of some fantasy or fairyland, one that allowed for the inclusion of a variety of plants including, a guidebook informed me, 'a huge clump of *Gunnera manicata*, a South American rhubarb which has leaves higher than eight foot'.

The sense of alienation I experienced was heightened by some of my fellow visitors. There were the two American girls who entered its doors at roughly the same time as me. They noted with open-mouthed reverence some of the items on display – the 10,000 books on the shelves of Duchess Eileen's library, their old-fashioned titles on display; the military heirlooms, complete with plumed helmets; the child's playroom; the bathroom placed – ingeniously 'at a later date' – within the building's round towers. Their sense of astonishment and wonder grew even greater when they reached the castle shop. All sorts of goodies were on display. Deerstalker hats. Tartan scarves. 'Wee bothies' courtesy of a firm called Glenshee pottery. I looked around for something that might place these 'wee bothies' in some kind of context, something that might explain why there were so few 'wee bothies' these days in the county of Sutherland, a short inscription that might provide a few reasons why it is the part of Scotland that seems most empty, but answer came there none. As far as I could see, there was no book on either its shelves or Duchess Eileen's library about the Highland Clearances or the role of the Sutherland family in those events. Instead, there were a few mock-Victorian guidebooks instructing readers on *How to be a Perfect Husband* or *How to be a Perfect Wife*.

'These look like fun,' I heard one American woman say to the other.

'Yes, they do,' the other responded, laughing as she read some obscure instructions that would ensure the woman was quiet, subservient and obedient to her man.

Sour and embittered Gaelic speaker that I am, I left them to the world of *Upstairs Downstairs* and *Downton Abbey*. I headed out to the museum that was in the castle garden.

Unaware of what lay within its doors, I staggered as I stepped within. The head of an elephant gazed out at me. A crocodile gaped. The antelopes mounted on the wall looked as if they had been tacked there by a sudden burst of gunfire. Attached to each one was a label noting the date and by whom it had been shot – invariably a member of the family that owned the castle. A rhino and elephant tail swished on their wooden boards, as if they had been left behind after an evening of S&M, part of the training involved in *How to be a Perfect Wife* or *Husband*. There was even a stuffed cheetah, which I was told had not been killed by anyone in the household.

'It's a gift,' the guard explained. 'Arrived here a few months ago.'

In short, the entire experience brought to mind the words of Harriet Beecher Stowe who, after visiting Dunrobin Castle in 1856, declared in a book entitled *Sunny Memories*:

> it is an almost sublime example of the benevolent employment of superior wealth and power in shortening the struggle of a civilisation and elevating in a few years a whole community to a point of education and material prosperity which, unassisted, they might never have obtained.

Hmmm.

And alongside this evidence of slaughter, there were many signs of displacement. The painted figurehead of the duke's yacht *Catania* loomed above me, stranded at the far end of the room. A Tibetan prayer bell, taken from a temple, stood silent and still a world away from where it had once been rung to call people to worship. Amidst all this exotica, there were small reminders of the local world nearby. Sometimes they appeared even more bizarre than the haul that had been brought from foreign parts. There were one or two wedges of tobacco they had apparently grown in the duke's Sutherland estate in the 1880s. With its long fibrous qualities, it resembled *calchas*, peat of a similar density one sometimes encountered on the peat bank, hard for a *tairsgeir* to cut through. There were occasions when, desperate for a cigarette in our teens, we would roll it up and place it either within a Rizla paper or a torn piece of newspaper. Suffice to say, it rarely made a satisfying smoke.

There were remembrances, too, of the most northerly coalfields in these islands, the industry which had once occurred in nearby Brora until it closed down for the final time in the mid-seventies. The exhibition also included a few samples from another unlikely event, the gold rush that in 1869 occurred at Baile an Òir ('Gold Town') in the Strath Kildonan area near the Helmsdale River. It was a location where I stopped during my travels, looking at the stream thundering down, wishing I had enough time to purchase prospecting equipment – such as a pan, a hand sieve, a small trowel and hand spade – to scour its torrents for the gold that still gleamed there. I comforted myself with the thought that its owners had seen greater prospects of wealth in the sheep that grazed nearby, the trout and salmon that also glinted in these waters. All that glimmered clearly was not gold.

Most incongruous – and yet most impressive – of all were all the Pictish standing stones that had been found within Sutherland. Etched on these, the Dunrobin Stone, the Colliesburn and Golspie Cross Slab Stones, were a few creatures that looked as if they were displaced from their normal habitats – a bear, a giant fish, a 'big cat'. There was even a stone with a creature someone helpfully described as a 'space invader' chiselled on its surface. This was not the only other-worldly monument to be found near Dunrobin Castle. Another is the 'Mannie', the 100-foot-tall statue dedicated to the 'social reformer' George Leveson-Gower, the first Duke of Sutherland, who displaced many of the people who lived in these parts during the nineteenth century, part of the era known throughout the north of Scotland as the Highland Clearances. He stands beside Ben

Bhraggie, one of the high peaks nearby, asserting his dominance over his former domain.

It is one of a few reminders of that power today. It can even be seen in the street names of Helmsdale, a township he created to take advantage of the herring fishing that became of great economic importance to the Highlands during that period. In a landscape dominated by peaks that have Gaelic titles, such as Creag Marail and Creag Bun-Uillidh, some of its streets bear names like Stafford Street, Lilleshall Street and Trentham Street, in honour of his lordship's estates elsewhere. Someone writing to the local newspaper the *Northern Times* declared – with perhaps a little dramatic exaggeration – that this was akin to towns in Germany bearing titles like Himmler Crescent and Goering Road. There would be others who might argue that the history of the area is a little more complex than this, that the intentions of his lordship were a little more benign than he is frequently given credit for. It is not a case for which I have much sympathy. Judging even by the shelves of their shop in which there can be found no balance or counterweight to a rather sentimental view of Highland history, there seems to be little awareness of the way in which both the family and the castle are viewed by many people in the north of Scotland even today. This even extends to their current plans to create a single-malt distillery at Dunrobin. Some of the suggestions for the brand name of the whisky they will eventually produce include 'Dun Robbin', 'Best Sellar', 'Croftburn' and the 'Burning Thatch'. The last three names are in tribute to the incident in which his lordship's one-time factor, the notorious Patrick Sellar, presided over 'the leaping flames of fire', as the Gaelic poet Derick Thomson described, when the thatched roof of a family called Sinclair in Strathnaver was set ablaze.* An old woman named Margaret MacKay was in the house at the time. Two years

* The poem also has the place name 'Srath Nabhair' in its title.

later in 1816, Sellar stood trial for this offence. He was found not guilty, but the verdict of both history and many of the ordinary people of the Highlands is at odds with this view.

Sutherland is also an area I have explored a little over the years, coming across its landscape first in the Gaelic poetry of Rob Donn MacKay and the English poet Norman MacCaig, and novels like Neil M. Gunn's *Butcher's Broom* and Iain Crichton Smith's *Consider the Lilies*, and then in historical works such as the writings of John Prebble – when I was a teenager – and later the works of historians such as Eric Richards and, especially, James Hunter. Each writer in his own way brought enlightenment, making me aware of some of the darker aspects of the history of my own people. It was my encounter with these works that made me alter – *beag air bheag* – my own attitude to the Gaelic language. For much of my adolescence, I regarded the tongue as backward and antiquated, something to be shaken off and forgotten rather than guarded and loved. It was when I approached my early twenties that my attitudes changed. I began to appreciate the language's subtleties, the differences between the attitudes of those who spoke it and those whose views were often constrained by possessing only one means of expressing themselves. I also came across other works, such as the writings of modern Gaelic poets like Sorley MacLean and – once again – Iain Crichton Smith, who convinced me that sometimes those fluent in English and Gaelic possessed an international quality that the largely Anglo-American viewpoint possessed by too many limited to one tongue missed. Gaelic had an echo that sometimes English on its own lacked, being the voice of those who had been dispossessed – of their own culture and, as the Mannie standing in front of Ben Bhraggie reminded us, land.

For many years I often glimpsed that landscape, its peaks and glens purple on the horizon as I stood anywhere on the east coast of the island of Lewis. Its geography was more visible than that of my imaginary Newfoundland. I could see the 'Old Bonadventures, Kettle Coves, Heart's Contents and Swift Currents' that might be there. I even knew the place names of real townships that existed on the edge of the Scottish mainland, locations like Lochinver, Kinlochbervie, Durness, Scourie and even – more to the south – Achiltibuie. Over the last few years I have sometimes driven round these townships, negotiating the dips and rises found along their coastline, the occasional brinks of cliffs falling towards the sea, marvelling at the long miles of emptiness I encounter, the deep coffee-shaded grey or red of moorland changing with the light of day or season, knowing that the main proof of life I encounter will be starlings, rising perhaps upwards like a freak wave, or crows, croaking and flapping in from some dark corner of the landscape where carrion has been found. There is little doubt what they will largely feed upon. Some notices I encounter remind me of their likely meal, warning – in German as well as English because of the assortment of tourists from that country driving round its roads:

Caution: Lambs On Road
Achtung: Lammer Auf Der Strasse

Or, of course, deer. Going out towards the village of Scourie, I was conscious at all times of the steep unpredictability of the roads in that part of Scotland – an endless squiggling trail of blind and vertiginous twists. Accompanying them, to complete my agony, herds of deer seemed to hang out like gangs of shy and inscrutable teenagers at every corner, grazing on each steep and unpredictable slope. Half of me was lost in the rapture of observing these shy and beautiful creatures. With each stag I glimpsed, I thought of the majesty of these Monarchs of the Glen; each fawn I watched, I thought of Bambi.

Another part of me, however, recalled the damage that can be created by these miraculous animals: the impact of their weight and strength against the body of a car, how their meat and muscle, gristle and fur can shatter glass and become laced within steel. From living in Uist I recalled too often the impact that such life can have on even the most sturdy of vehicles, the wreckage that can ensue. It seemed to be all of a piece with the way human life had been affected in this, one of the most empty and desolate stretches of landscape in northern Europe – the bare existence that people could find here slipping from their fingers, condemned to years of hunger and decay.

There was less sense of this travelling from the east coastline, driving along the fast-flowing torrent of the Helmsdale River and towards Strathnaver. I occasionally stopped en route, noting that much of the Strath of Kildonan appeared a green and fertile place, suitable for farmers to plough and trees to root within. In any other part of the country, there would be a cluster of new houses, edging their way into the emptiness that ranges across the heart of Sutherland. It is, as James Hunter reminds us,

> twenty-first-century Britain's empty quarter. England's average population density is 413 people per square kilometre. Scotland's is 68. But each square kilometre in Sutherland, a county of roughly the same size as Norfolk or Northumberland, contains on average two people: and since most Sutherland residents live in coastal areas, much of the district's extensive interior is practically uninhabited.

It is an emptiness about which I heard a great deal during my childhood in the sixties. The county of Sutherland contained a rather high percentage of Scotland's 1.7 million hectares of moorland, yet it had not always been so. One of the main reasons for the area's emptiness was even contained in one of the rare books that used to circulate around the crofting villages of the Isle of Lewis in my early years. It contained the predictions of a gentleman called either the

Brahan Seer or Coinneach Odhar, Kenneth (the sallow-skinned) Mackenzie, who was apparently born in the village of Baile na Cille in Uig on the western coastline of my native island some time near the beginning of the seventeenth century. Many of his prophecies were noted and recalled. Employing a stone which he had allegedly been gifted by the ghost of a Danish princess his mother had met one night in a graveyard, he foresaw the Battle of Culloden near Inverness in 1746, when Jacobite forces were defeated by Hanoverian troops. Standing on that stretch of moorland, he was said to have declared:

> Oh, Drumossie, thy bleak moor shall, ere many generations have passed away, be stained with the best blood of the Highlands. Glad am I that I will not see the day, for it will be a fearful period; heads will be lopped off by the score and no mercy will be shown or quarter given on either side.

His other words include the unlikely notion that Inverness County Council was responsible for the outbreak of the Second World War. (Apparently, a major calamity would befall the planet after the third bridge was built over the River Ness. A short time after they decided to do this, the Second World War took place.) Yet the forecast that I found most memorable was the one that related to the emptiness of Sutherland. He had said that over much of the Highlands, 'the sheep would eat the men'.

And so, in an odd way throughout this empty landscape, this transpired. As was noted by another seer or prophet, Karl Marx, in his article on 'The Duchess of Sutherland and Slavery' published in January 1853, the Countess of Sutherland attempted to improve the economy of the Highlands by being 'determined to transform the whole tract of country into sheep-walks'. They nibbled and chewed their way through the emptied townships of places like Rossal, gnawing the occasional stalk of grass in fields where beremeal had once been planted, sheltering too behind the ruined walls of houses in locations similar to Ascoilemore in the middle of Strathbrora. They bleated where children had once run and played in the likes of Kinbrace, Suisgil, Torrish. My car had to come to a halt on a few occasions when a small flock of sheep crossed the road, making its way in the direction of the river. So much, so familiar, as Karl Marx himself notes in that same document:

> The process of *clearing estates*, which, in Scotland, we have just now described, was carried out in England in the 16th, 17th, and 18th centuries. Thomas Morus already complains of it in the beginning of the sixteenth century. It was performed in Scotland in the beginning of the 19th, and in Ireland it is now in full progress. The noble Viscount Palmerston, too, some years ago cleared of men his property in Ireland, exactly in the manner described above.

Yet it seems to me that, for all that these events occurred elsewhere, they left a more lasting legacy in places like the Highlands and Islands of Scotland and Ireland. Both land and language are politicised there to a much greater degree than south of the border. There are reasons for that. The Highland Clearances came later than their English equivalent, bringing them closer both to people's imagination and to their understanding. Then there was the scale within a limited area. According to information posted in the Strathnaver Museum, the Clearances involved the emptying of 39 townships before 1815 with 212 families removed and 1,000 acres of arable land cleared of its population. In 1819–20 alone, around 1,000 families, with an average of five persons each, were forcibly sent from their homes. In addition to this second wave of the Highland Clearances, which lasted from 1800 to 1850, there had also been a previous surge in and around the 1760s. These actions involved the emptying of a distinctive language and culture from the areas affected, replacing the Gaelic tongue with either English or – more frequently – the Scots speech of many of the Lowland shepherds brought to these glens to work and help graze sheep. This allowed the blaming of a historical wrong on 'the other', largely the English, though any real examination of the facts would lead one to the conclusion that two of the principal villains in the Clearances – the unspeakable Patrick Sellar and James Loch – were, in fact, Scots. Those displaced by the Enclosure Act in England had no such alien overlords to hold responsible for such actions. They knew there was no great evidence that there was anyone other than themselves, the English, to blame.

This has had its political effects on both the internal and external politics of Scotland even to the present day. Most evidently, it allowed some of those who were 'cleared' or indeed 'remained' to blame those south of the border for the sheer brutality of the Clearances. Within the country, however, the resentment felt by many in the Central Belt at the arrival of Highland and Irish immigrants in cities like Glasgow and Edinburgh is, perhaps, one of the principal reasons why there is still hostility towards 'teuchters' and their language. It has involved a certain rewriting of history, a rather remarkable obsession among certain Lowlanders about where and when Gaelic was spoken. The letters column of the *Scotsman* newspaper is full of such demarcation disputes, often focusing on the great expense that is allegedly involved when signs for Scottish railway stations are replaced with their bilingual equivalent. Individuals point out with great certainty that their area was one where Pictish was the dominant language or where Angles settled. They bear swords in the service of imaginary ancestors and the ghosts of former tongues. Few are aware that Gaelic stretched as far as Upper Deeside until the mid-nineteenth century or that the Bibles read from Highland pulpits were written in the Perthshire variety of the tongue. The differences between the moorlands and languages of the Highlands and those of the gentler lowlands of the Scottish Central Belt are not as sharp and stark as sometimes portrayed. Deep peat, jutting rocks and sphagnum moss are sometimes found to the south; fertile land is occasionally discovered to the south. The same is true of the tongue that was heard in these areas.

The Highland Clearances and other similar events also crammed too great a percentage of the population of Scotland into what is – from west to east – a particularly narrow stretch of land, in tenement buildings and high-rise flats that stand within a rather tightly buckled Central Belt. As a result of this, the interests and politics of the north have often been neglected. For all that its emptiness is often romanticised and viewed as 'wilderness' by those who come to visit, for those who live there, there is isolation and desolation that often seems unnecessary and avoidable, and could be addressed if there was a modicum of political will or interest. The moors of Sutherland, particularly near that county's eastern coastline, are not as barren and infertile as they have often been claimed to be, and are capable of supporting a higher population than they currently do. Despite the propaganda of Sellars and his ilk, the reason that places like the Strath of Kildonan were chosen to be 'cleared' was not because they were infertile and unproductive, but because they possessed the promise of prosperity, offering, within the context of the expanse of Sutherland as a whole, rich grazings for the sheep.

One further reason for the richer, deeper witness to these events than, say, those that followed the Enclosure Act in England is that the glens of Sutherland also contained eloquent, articulate individuals whose voices still persist. They included men like Donald Sage, son of Alexander Sage, who was minister in Kildonan in 1813. He gives testimony to some of the lies generations have been told about the Sutherland landscape in his memoirs, unavailable until 1889. He describes the woodlands of 'black willow, oak, aspen, alder, and wild gean [or cherry], the mountain ash or rowan, the black flowering thorn and the birch tree' to be found there. He tells, too, of the amount of animals sold in the Strath of Kildonan in April 1791, enumerating the sheer quantity of cattle, sheep, horses and goats, and the prices – horses from £4 to £6 sterling, cattle from 50 shillings to £4 and 10 shillings – obtained for them at market. As the historian James Hunter remarks, these are, in relative terms, 'better than those fetched in Scottish livestock markets at the turn of the twenty-first century'.

This contrasts with Edinburgh lawyer and sheep farmer Patrick Sellar's words. He described it as 'a country of sloth and idleness' where people were raised in 'poverty' and 'beggary'. This, of course, was part of his justification for the 'improvement' he and his kind were about to bring to the Sutherland estate, replacing both the native sheep and the 'aborigines' who had lived there for centuries with the Cheviot sheep. Not content with removing both man and flock, he also set about destroying much of the wildlife that existed there, doing his utmost to ensure that birds like the sea eagle and red kite were on their flight to extinction within these islands. Behaving in a manner that is yet to disappear in the grouse moors still found in the British Isles today, the collection of birds and animals destroyed by Sellar and his allies between August 1819 and March 1820 bore a great similarity to the host of more exotic creatures mounted and displayed in Dunrobin Museum. They included: '112 fully grown eagles [both sea and golden eagles], 18 young eagles, 211 foxes, 317 wildcats, [pine] martens and polecats, 516 ravens, 218 hawks [including, presumably, red kites and hen harriers], 1,183 carrion crows and magpies, and 570 rooks and jackdaws'.

The list provides further evidence of the manner in which land was laid to waste by the duke and his cronies. It is this devastation that lies behind the splendour and wealth of the Italianate garden at Dunrobin Castle, the displacement of human and animal life which occurred during this period concealed behind its richness and fertility. For all that the book *Gloomy Memories* may not be found on the shelves of the shop of Dunrobin Castle, one can find proof of that brutality in the testimony of an individual whose memorial can – like the Mannie – be found within the county of Sutherland. Donald Macleod's monument does not dominate the landscape in the way George Leveson-Gower's high tower does. Instead, it lies on low ground where, rather ironically, it offered shelter for a pair of ewes when I drove towards it one afternoon. They cropped the grass nonchalantly, unaware of the inscription above their heads.

> In Memory of Donald Macleod, Stonemason, who witnessed the destruction of Rossal in 1814 and wrote 'Gloomy Memories'

My first encounter with this man was in Iain Crichton Smith's *Consider The Lilies*. Seen in this work as a rebellious freethinker, Donald Macleod is portrayed as being on bad terms with both the church and several of his neighbours, including Mrs Scott, the pious widow who is the central character of Crichton Smith's book. It is her journey that provides the focus of the novel, one that conveys the manner in which she comes to the slow and uncomfortable realisation that she can no longer rely on the factor, Patrick Sellar, nor the local minister when she is threatened with eviction. Instead, it is the atheistic Macleod who provides her with her sole comfort, consolation and support.

Crichton Smith does not move too far from reality in the way the fictional Donald Macleod is portrayed in his book. The real figure was outspoken and unconventional for the time, displaying this tendency particularly in the way he gives witness to the events that occurred in 1814, known as 'The Year of the Burnings'. He provides his greatest testimony of these times in the book *Gloomy Memories*, his own rejoinder to Harriet Beecher Stowe's work, *Sunny Memories*, which she completed after her visit to Dunrobin. He describes how there were:

> 250 blazing houses [one night in that community]. Many of the owners were my relatives and all of whom I personally knew; but whose present condition, whether in or out of the flames, I could not tell. The fire lasted six days, till the whole of the dwellings were reduced to ashes or smoking ruins.

Much of his writing was printed initially in the *Edinburgh Evening Chronicle*, which published twenty-one of his letters detailing the evils that had been done to the Highlands at this time. It had a great effect on public opinion in the Lowlands of Scotland, which began to recognise the wrongs that had been done in the country's far north. However, this also had a great effect on Macleod's personal life. His family were removed from various homes throughout the Highlands. His wife sank into madness. Eventually, he had to travel south to Edinburgh and from there to Canada, where he wrote *Gloomy Memories*. There is no doubt, too, that it had a great effect on his relationships with his neighbours; several of them would have distrusted him because of his radical, unconventional ideas. It is this that lies behind his clash with the character Mrs Scott, the archetypal Highland woman who lives near his home. She is portrayed as being loyal to the church, never questioning the minister, despite the fact that he is clearly acting in the landlord's interest and not in hers.

Near Ardgay in Sutherland, there is a building I visited a few years ago which shows that Mrs Scott, while clearly invented, was not a fiction. Set among moorland, a small clump of trees around its white walls, Croick Church offers its own testimony to the cruelties of that age. In 1843, the people of nearby Glen Calvie were cleared from their homes and summoned to the town of Tain for a 'friendly discussion' by their factor, James Gillanders. The 'discussion' clearly became less 'cordial' when the gentleman, acting, perhaps, on behalf of his landlord, Major Charles Robertson of Kindeace, handed his 200 tenants in that area an immediate writ for their removal. While some obtained warmth and shelter from relations and friends nearby, there were others who obtained no such shelter and kindness. It was only with the factor's kindness that the women were allowed to stay in Croick churchyard, while their husbands, sons and brothers travelled to find both new homes and new employment.

Their stay was recorded in two ways. A correspondent from *The Times* happened to be on hand to cover the events. He noted that 'through the actions of factors in the lonely glens, hundreds of people and generally industrious peasants have been driven from their means of support to become wanderers and starving beggars – a brave and valuable population lost'.

The other method was a little less transportable than the pages of a newspaper. Engraved in the glass windows of the church are a series of inscriptions. They include not only the names of those dispossessed, such as 'C. Chalmers', 'Ann McAlister', 'John Ross, Shepherd, parish of Ardgay', but also inscriptions like 'Glen Calvie the wicked generation', a Biblical quotation from Matthew 16, verse 4. Their choice of text sums up the violence that was often done to the mindset of those who were displaced. They accepted the blame for the wrongs that others had done to them, the crimes and injustices that those who were in authority had inflicted on their lives.

And in this, they were not unique.

CHAPTER ELEVEN

Musta Turve (Finnish) – Black Peat

It is not only moorland that man has sought to 'improve' over the centuries.

Climb the steps of the building where in my childhood I was most likely to catch a glimpse of Paradise on Earth (in reality a parish of St Johns, Newfoundland, with a population of 17,695 in the Canadian 2011 census) and one can see evidence of this. Look west from the top of the Butt of Lewis lighthouse and the viewer can only witness the breadth of the Atlantic with all its tempestuous waves, perhaps play a game of tiptoes and imagine spotting Old Bonadventure, Swift Current and Kettle Cove. However, look east and downwards, close to the walls of the old lighthouse-keeper's quarters, and it is possible to distinguish a different kind of wave pattern on the surface of the land. It resembles the floor of a waltzer, one of those fairground rides we might have encountered at some time in our childhood if a travelling fair had ever deigned to visit the town of Stornoway. We would skip and run over its undulations, imagining we were spinning and turning, occasionally toppling on the short sparse grass below our feet.

How this feature had been introduced to our childhood landscape had not, however, been quite so much fun. This stretch of land, near the village of Eoropie, was among the many early efforts to cultivate the moorland of the Isle of Lewis and elsewhere on the coast of the Highlands and Islands. Called by various names – *feannagan* in Gaelic, run-rig or the rather ironically titled lazy beds in

English – this sweeping, rolling stretch of ground was the first attempt by settlers in these parts to transform the rough moorland to which they had fallen heir. They would plough up the barren soil with the old *cas-chrom* (foot plough), creating a trench in which rotting seaweed and manure were placed. Within these furrows, they would plant beremeal and other crops, trying to ensure there would be food enough in this harsh environment for their families to eat. After this, there would only be the smack and sweep of salt to contend with, carried on the spray that washes over *Rubha Robhanais*, the Gaelic name for the Butt of Lewis. For all my childhood pleasures here on this curving part of the island's shoreline, there is no doubt it would have been a hard and bitter life.

Yet improving the moorland did not occur only in the places at the edge of our continent. It also occurred elsewhere. Sometimes, too, it was linked with not only improving the moor, but also an attempt to modify the nature of the poor – the 'wicked generations' – that lived throughout the nations of Europe. One vivid example of this was written about by Dutch author Suzanna Jansen in her book *Het pauperparadijs* (*The Paupers' Paradise*). A former secretary, market researcher and journalist based in Moscow, and at one time employed in a ballet academy until she was injured, she brought her intelligence and insight to an aspect of the history of the Netherlands that involved her own family background. It centred on a similar, though smaller, community to the one Elleke Bal came from, Veenhuizen in Drenthe in the northwest of the country. Inhabited today by some 800 souls, it actually had its beginnings in the dreams and visions of an army officer, General Johannes van den Bosch, who later became governor-general of the Dutch East Indies and minister for the colonies in the Dutch government in the nineteenth century. A paternalistic figure, both during his time abroad and at home in the Netherlands, Johannes van den Bosch set up the Society of

Humanitarianism – also known as Society of Benevolence – in 1818. His task was revolutionary. In the new kingdom of the Netherlands, formed after the collapse of the First French Empire, there was a great deal of poverty and want. The Napoleonic Wars had devastated the country's economy and its people were hungry and in need of work. Around half the population of Leiden were given famine relief, often in the form of soup kitchens. Rotterdam had around 800 beggars roaming its streets. In Amsterdam, children crammed the orphanage, either orphaned or abandoned by their parents. It was a situation that Van den Bosch hoped to begin to remedy by obtaining a penny every week from some of the 20,000 richer citizens he enrolled as members of his scheme. All this would contribute to the education of paupers and the restoration of the homeland. It would also solve two separate problems. Looking towards Drenthe, he saw that 70 per cent of the countryside – particularly the *mohrs* (moors) – were not farmed or used for the benefit of humanity. He believed, too, that there was a similar percentage of the Dutch people of whom something comparable might be said.

He started small. There were 52 families initially in the scheme, brought from the cities to a place near Steenwijk in the north-eastern part of the Netherlands, an area bordering Drenthe and its moorlands. Poor families were given a farm there, encouraged, perhaps, to create new communities, free of the troubles of the old. They would be taught to cultivate the land and grow food, much as my ancestors had probably done when they had dug out 'lazy beds' on the coastline of the island of Lewis. As Suzanna pointed out, they lived under tight surveillance, taught by discipline and punishment to be good citizens rather than the social outcasts of the cities they had left. The children would also be taught to read and write, an essential element of the society that Van den Bosch wished to create.

However, the general's vision was much wider and more ambitious than this. Wishing to eradicate poverty in his country, he grew impatient. He wanted to re-educate 40,000 people, preferably in a maximum of 12 years. As a result, he decided to set up large institutions in the moors. This was cheaper and more economic than elsewhere. The land was of little value, for all that this might be one day altered by the labours of men. There was a great deal of space to build. There was even an institution, for instance, in Ommerschans, an old fort not far from Hoogeveen, the former haunt of Vincent van Gogh, which was marked down on the map as *Locus Deserta Atque ob Multos Paludes Invia* (deserted and impenetrable place of many swamps). This institution was something like a penal colony, brought into being with the intention of transforming the nature and character of individuals who had settled there as well as cultivating the moorland nearby. Other institutions of similar ilk followed. Of them, Veenhuizen was the largest, consisting of three institutions with room for 2,000 to 3,000 individuals, even more if an emergency arose.

There were difficulties in all of this. Van den Bosch discovered that those among the poor who were capable of working refused to go to these institutions. They were too far from their friends and families in the larger cities and towns. Their reputation as repressive places became known farther and farther afield. They became locations where the poor were forced to undertake labours not unlike those described in Robert Garioch's Scots poem 'Sisyphus', where:

> Bumpity doon in the corrie gaed whudderan the pitiless whun stane.
>
> Sisyphus, pechan an sweitan, disjaskit, forfeuchan and broun'd aff,
>
> sat on the heather a hanlawhile, houpan the Boss didna spy him.

The general dealt with the infamy of these places in several ways, convincing King William I that the orphans of the Dutch cities would be better off in the fresh air and open ground of locations like Veenhuizen than the cities from which they came – and much less expensive to keep there too. Afterwards it was decided that all the 'city boarders' over the age of six would be sent there, too. Later still it became the home of the sick, the elderly, the disabled. Their accommodation was divided. On the outside, there were the 'deserving poor', some 1,400 families who stayed in one room but could enter and leave the building through their own door. In the building's interior, within a dormitory, there slept the beggars, vagrants, prostitutes, those whose honesty could not be vouchsafed or relied upon. Within there, the sexes slept apart, stepping out in the early morning to a courtyard that was locked and closed. There were similar arrangements for the orphans, their every movement supervised. In her book, Suzanna writes of seeing this, noting how, after walking round the building, which has now been transformed into a museum:

> it's the sleeping berths that cause me to slow my pace. A row of six white alcoves made of woven steel bands, 90 cm wide, 1 metre 85 long. One of the alcoves is invitingly open. As I look into it, two ladies walk over towards me. Above the clicking of their heels I hear them talking about project estimates and programme profiles.
>
> 'Oh, how sweet,' one of them lets slip spontaneously on seeing the berths. 'Look here, these must have been for the small criminals.'

It is a remark that Suzanna responds to with understandable anger, feeling the blood come to her cheeks when she listens to their words.

The three institutions built at Veenhuizen first became part of the collective experience of Suzanna Jansen's family four generations ago. The earliest of her ancestors to step in through its doors was an old soldier, one of many the general used to supervise the institutions. As Suzanna told me, it was a role that later led to a twist of fate. The old soldier's daughter fell in love with a pauper's son, who was among those being guarded. As a result, she lost the status of being a guard's daughter, for all that even this existence was often miserable and poor, and instead began living as an inmate in the building they called the Third Institution for a total of 33 years of her life. It was there her children were born, not all of whom made it through their first years. Her youngest daughter did survive: Suzanna's great-grandmother, brought up in that institution.

It was only when the rules relating to their prison confinement changed in 1859 that she left with her offspring to Amsterdam. ('In reality not that much different from before,' Suzanna said. 'Even the poverty in Amsterdam was just as grim at that time as it had been in Veenhuizen.') This meant that there were no poor families within its walls any more. Instead there were only beggars, with the designation of Veenhuizen changed to a state labour institution. Yet the name of the place carried its own stigma. The family would lie even in official documents about their Veenhuizen background, dodging the shame and humiliation associated with it, aware that even the mention of its name made it difficult for them to find work in Amsterdam. For all that they had passed through the doors of the building, the family's destitution proved a further insurmountable trap. Suzanna's great-grandmother married a man who, later in his life, could not stand the poverty and the loss of his children any more. The wheel turned again. He started wandering and begging, until this resulted in his being charged by the authorities for vagrancy. As a result, he was sent to Veenhuizen in 1900.

In her book, Suzanna provides a moving description of an individual which she noticed displayed on a board in the museum:

> The personal description for convict number 5374 states that he is 1 metre 50 tall. The width of his head is 15.6 cm, its length 18.9. His facial pigment is whitish yellow with a transparent tinge, the ridge of the nose straight, the left eye mid-blue. The sheet also gives the length of his sentence: three years, expiring on 5 November 1903.
>
> For safety's sake, four fingerprints were taken of his right hand. Then 5374 was photographed, in profile and full face, his number always in view. He stares at me, inmate number 5374, with a washed-out expression. Passive and at the same time slightly startled; he has never stood in front of a camera before.
>
> Finally, 5374's personal description card gives his name: Harmen Keijzer. And his age: forty-eight. I stare at the man in the photo as he stares at me. This is not just any vagrant; this shabby figure is my great-grandfather.

And so the family history – even until the third generation – returned in a circle to this place. It was one that – even in the way Suzanna discovered it – is marked by a sense of the family's shame, declaring that:

> I discovered this history when I found a prayer card for my great-grandmother, who was supposed to have been born in Norg – a village in Drenthe. My mother told me this great-grandmother was raised as a Protestant but fell in love with a Catholic. For this reason she lost the support of her family and community, and had to move to Amsterdam to find work. This was why the family became poor. Curious, I asked for her birth certificate – to find out the address where she was born, so I could see the house that she had lost. But

> the birth certificate said something else: it mentioned she had been born in 'the 3rd institution of Veenhuizen.' I had never heard of Veenhuizen, but understood that being born in an institution, and the story I had heard before, did not match. That is why I started to research. The family story turned out the be a smokescreen for the shame of having been in Veenhuizen. It was necessary to lie about this, as otherwise, you would not get work.

There are some who might claim that there were benefits to life in places like Veenhuizen, that the children learned to read and write in a manner denied to their contemporaries elsewhere. It is not a view supported, however, by Suzanna.

> I cannot say the idea was a success. Some people who lived in the earlier scheme, who lived near Steenwijk were better off, staying together as a family, all on a smaller scale. They were not forced to lie about their background. But I cannot say the same of Veenhuizen. It was very hard to ever leave Veenhuizen once you ended up there, often as a last resort against hunger. The rules for leaving were very strict and constantly changing. And those in charge of people's education, often themselves hardly had any education. Families were torn apart, with children in a separate institution – only seeing their parents on Sunday afternoon, guarded and watched over all the time. Nobody actually gained a profession from their time there. The stigma that people from Veenhuizen possessed made it almost impossible to find a job outside. It contributed largely to the feeling they were not worth anything, something which continued for generations, even affecting my own views and feelings. So many 'lost' lives in the moors.

The statistics underline Suzanna's words. She notes that – at the time – 3 per cent of the Dutch population were in establishments like Veenhuizen, 'cleared' from their homes in an urban environment and brought to what has been termed the 'Dutch Siberia' in a strange and ironic subversion of the events that occurred in my country. According to demographic evidence, this means that around 1 million of the approximately 17 million people that make up the present population of the Netherlands had ancestors in these institutions. Many of them will have read Suzanna's book; 300,000 copies were sold, a bestseller in the Netherlands. However, in addition to all her previous occupations, she has also been involved as a dramaturg in theatre, performing an important role in the production of *Het pauperparadijs* that took part in Veenhuizen in the summer of 2015, performed in the only building from 1823 that still stands. On 40 occasions during that period, a thousand people came to the community to watch this play, one that not only revealed events of that time but also provoked debate about attitudes to the poor that still exist in contemporary society.

Veenhuizen was not the only part of Europe with a theatrical atmosphere that summer. It was also true of Aarhus, the Danish city on the peninsula of Jutland, which I visited in 2016. Its streets were crowded with young and old when I arrived there, all intent on marking the season with a riot of revelry that involving young men singing the lyrics of Johnny's Cash song 'I Walk the Line' while leaping into the narrow river that gives the city its name.

Perhaps it was the rainbow that arced above a building in the city that engendered this feeling – one that I am aware was exceptionally lively, especially by Danish standards. An array of lights swirled above Aros – an art gallery with a similar name to the arts venue in Portree in the Isle of Skye, which I know well – beckoning people to its doors. Known as the rainbow panorama, it is the work of Icelandic artist Olafur Eliasson. 'Here be giants,' the spectrum of colours seemed to announce, drawing in visitors arriving in this part of Jutland on their cruise ships on even the most dull and grey of days. 'Here be wonders.' Inside its walls there was a

5-metre-tall human sculpture, *Boy*, by Australian artist Ron Mueck. I had seen some of this sculptor's work before in Aberdeen and was anxious to see more. There was a gargantuan *Wild Man*, a *Spooning Couple* and a giant *Baby*. In each of these, every wrinkle and crease, each follicle of hair seemed to be in place.

But I had no time to visit *Boy*. Instead I was off to see a couple of smaller human figures, both with considerably less clothing than the pair of shorts worn by this particular example of Mueck's work. On the leafy edges of Aarhus is the last resting place of the Grauballe Man, dug out of a peat bog in a district of that name. Nowadays he is found within a darkened glass case in the modern structure of the Moesgaard Museum. Within its walls, there is a wide and shimmering stairway illustrating the evolution of mankind. Guarding the visitor as they step upon it are seven 'hominims', designed to show the beginnings of mankind. A naked female leans upon a stick. Another individual – dark-skinned and with a shock of black hair – rests his spear upon his shoulder. A naked man juts his stomach in the direction of new arrivals. None of these figures, however, are as impressive as the Grauballe Man in the darkness of the building's lower floors. Discovered in the part of Jutland that now provides him with his identity, he was once thought by some to be a relatively recent resident of the district, specifically 'Red Christian', a peat cutter in the parish who had gone missing several years before, shortly after a visit to a local inn. In the spirit of villagers everywhere, several even came forward as reliable witnesses to this fact.

However, carbon-dating later revealed this was not the individual locals thought it was. Instead, he was found to date from the third century BC, last walking over the turf in the early Germanic Iron Age. Nevertheless there is much about him that might make people suspect he was around at a later date. These include the mat of hair on his head, the whorl of his fingerprints, even the fact that the condition of his hands allowed those who examined him to decide that he had never done any harsh, physical work, unlike the vast majority of those who would have surrounded him at one time. Adding to the story, there was the damage inflicted upon his head that, according to Heaney in the poem he wrote about the figure, 'left him bruised as a forceps baby', and the slash on his throat. Together, there was enough evidence to create a background story for this man who now lies stretched out and propped up for people to peer at in his case in Moesgaard Museum, one with lighting almost as shadowy and subdued as the peat bog from which he had emerged.

And then there is the Tollund Man. Naked apart from a skin cap and a leather belt around his waistline, this timeless, ageless body lies these days in a museum in Silkeborg, some 44 kilometres out of Aarhus. It took a short time to reach him, my morning train rolling through the Danish landscape as I took note of the anonymous railway stations, red-roofed housing estates and green farmland en route. Somewhere on that journey, I was surprised by the sight of a short stretch of moorland, a round, purple-crested hill glimpsed through the train window. Its shade and contours came as a small and pleasant shock to me, reminding me of the bloom of heather in the Scottish moors during August. Upon my arrival in Silkeborg, and after the inevitable loop and curl around its streets when I took the wrong direction, I finally, behind the doors of a large yellow building, found where Tollund Man lay.

He had odd bedfellows in his new residence. In the room beside him were the odds and ends churned up when a new motorway had been opened up to Silkeborg. They included Mesolithic axes and pointed weapons found in the mud and sludge of Lake Silkeborg Langsø; a Bronze Age vessel with a crust of brown food; a neck ring with a Thor's hammer pendant found in women's graves in Harup from the Viking Age; even – and herein hangs a tale – a hoard of empty condom boxes from the sixties, dug out from the Nordskoven forest. Nearby, too, was the figure they called the Elling Woman, who in June 1938 was discovered less than 100 metres from where the Tollund Man would be found some 12 years later. Wrapped in capes fashioned from calf and sheepskin, she is believed to have lain there since between 320 and 210 BC, the abrasions around her neck suggesting that she had been hanged before being placed in the bog.

I passed over much of this with probably less curiosity than I should have. Largely covered by a blanket of turf, Elling Woman looked much less human than the pictures I had seen of other bog bodies. I was not surprised to discover that the farmer who had first come across her suspected she was an animal. I thought her head looked like a rust-coloured stone. I was also conscious that I had made this journey for one reason only – to see the old gentleman stretched out in these parts. Eventually, after I had spent a little time examining the material on display outside, I stepped into his lair. Walled off by both a short partition and darkness, he lay sleeping there.

He had something in common with Mueck's work. Much of him is an artificial creation; his body, caught in a foetal position, was deemed too disturbing for public view and allowed to decay soon after its discovery in May 1950. The face is, however, real and authentic, for all that it had shrunk by about 12 per cent due to being preserved for a year in a concoction including formaldehyde, heated wax and paraffin. His expression was remarkably unaffected by this treatment. In the dim light, it is apparent that even after all these centuries lying in the peat bog, every wrinkle and crease is in place; so too were a few follicles of hair, especially the light growth of stubble on his chin. What was different was that, unlike the artist's glass-fibre creations, his eyes were almost shut, his smile quiet and benign. There were none of the startled expressions with which Mueck's giant sculptures viewed the world. Instead, he surveyed its madness calmly, protected – or so it seemed – by the peace of his 2,000-year-old dreams. It is this expression that has allowed the poetic imagination to spark into life. The American poet William Carlos Williams in 'The Smiling Dane' allows us to think of his executioners, faced with that smile upon his face, aware that they would 'grimace' when dealing with the reality of his death. In 'Punishment', Heaney transforms him into a Christ-like figure, the dignity of his passing a way of understanding those dead through the political violence of the Irish's poet's lifetime in Northern Ireland as well as those sacrificed in his own era. It is a way of looking at him that was at least partly inspired by P. V. Glob's words in *The Bog People*, written back in 1965:

> It is the dead man's lightly-closed eyes and half-closed lips, however, that give this unique face its distinctive expression, and call compellingly to mind the words of the world's oldest heroic epic, *Gilgamesh,* 'the dead and the sleeping, how they resemble one another'.

This smile is all the more startling because of the violent way he is said to have died. Half drowned, half throttled, he is alleged to have been thrown into the bog weighed down by a bough of rotting wood. The reasons for this act are, of course, open to conjecture. The Roman historian Cornelius Tacitus in Germanicus round about AD 96 described the 'bog bodies' as *corpores infames* (infamous bodies), claiming the manner of their deaths told much about the offence they were said to have committed:

> Traitors and deserters are hanged on trees; cowards, shirkers and sodomites are pressed down under a wicker hurdle into the slimy mud of the bog. The distinction in the punishment is based on the idea that offenders against the state

> should be made a public example of, whereas deeds of shame should be buried out of men's sight.

While some of the writings of Tacitus have been dismissed as second-hand accounts of what was happening at the time, there is also evidence to suggest that there is truth in this particular claim. This includes the fact that those who were under suspicion of homosexual offences were occasionally treated in this way until relatively recent times. A purge of the gay population took place in the Netherlands in or around 1730, a trial of terror that began during June of that year when seven young persons were charged with 'the detestable Sin of Sodomy', a form of sexual activity formerly unknown in these parts 'and confined to the South Side of the Alps'. We are told that during this period:

> At least 60 men were sentenced to death. For example: In Amsterdam, Pieter Marteyn Janes Sohn and Johannes Keep, decorator, were strangled and burnt, 24 June 1730; Maurits van Eeden, house servant, and Cornelis Boes, age 18, Keep's servant, were each immersed alive in a barrel of water and drowned, 24 June; Laurens Hospuijn, Chief of Detectives in the Navy, was strangled and thrown into the water with a 100-pound weight, 16 September. At The Hague, Frans Verheyden, Cornelis Wassenaar, milkman, Pieter Styn, embroiderer of coats, Dirk van Royen, and Herman Mouilliont, servant, were hanged and afterwards thrown into the sea at Scheveningen with 50-pound weights, 12 June; Pieter van der Hal, grain carrier, Adriaen Kuyleman, glove launderer, David Muntslager, agent, and Willem la Feber, tavern keeper, were hanged and thrown into the sea with 100-pound weights, 21 July. In Kampen: Jan Westhoff and Steven Klok, soldiers, were strangled on the scaffold and buried under the gallows, 29 June. In Rotterdam, Leendert de Haas, age 60, candlemaker, Casper Schroder, distiller, Huibert v. Borselen, gentleman's servant, were strangled, burnt, and their ashes

> carried in an ash cart out of the city and then by ship to the sea and thrown overboard, 17 July. And at Zuidhorn, at least 22 men were executed on 24 September 1731, including Gerrit Loer, age 48, farmer; Hendrik Berents, age 32; Jan Berents, age 19 — all scorched while alive and then strangled and burnt to ashes; 12 others aged 20-45 were strangled and burnt; and eight youths aged 16–19 were strangled and burnt, including Jan Ides, age 18, who said upon hearing his sentence: 'I forgive you for the sin which you have committed against me.'*

Clearly this chimes with some of the words of Tacitus, with the corpses of these individuals being either burnt into ashes or plunged in water. It may be that the drowning of individuals for homosexuality has a long connection to the past.

It is not, however, the ritualistic deaths of the Bog Bodies that Seamus Heaney chooses to highlight in 'Punishment', his poem about the Windeby Girl, found across the Danish border in the north of Germany. He portrays her head as having been shaved. She has also been stripped naked before being thrown into the bog. Making an imaginative leap, he links her to other women who have been treated that way in the past – those punished by the French Resistance for collaborating with the Germans, the teenage women from his own Catholic Irish background who, during the Troubles, had been humiliated for going out with British soldiers. This notion of women being treated in this way for an adulterous relationship outside their marriage or their 'tribe'

* Rictor Norton (Ed.), 'Newspaper Reports, The Dutch Purge of Homosexuals, 1730', *Homosexuality in Eighteenth-Century England: A Sourcebook*. Updated 13 September 2000; reorganised and expanded 21 July 2002; updated 17 November 2011 http://www.rictornorton.co.uk/eighteen/1730news.htm.

is found again in Tacitus, who, recording the habits of the northern tribes outside Rome, claims:

> A guilty wife is summarily punished by her husband. He cuts off her hair, strips her naked, and in the presence of kinsmen turns her out of his house.

There is, however, one problem with this. Despite the way in which Seamus Heaney takes time to write about her 'naked front' with her nipples being likened to 'amber beads', her 'shaved head' being that of an 'adulteress', it has been established recently that the Windeby Girl – which I saw a number of years ago in Schleswig on the Baltic coast of modern Germany – was, in fact, male. A similar issue arose with the Weerdringe Couple found in a peat bog in Assen in the Netherlands. Again these two individuals, once believed to be a male and female, have now been discovered to be a pair of men.

These would seem to give credence to an unlikely authority on bog bodies – the notorious and reprehensible figure of SS Reichsführer Heinrich Himmler. He expressed his belief that many ritually sacrificed in peat bogs were homosexual in a meeting in which he addressed the Waffen-SS in 1937. In this, he announced:

> [Our ancestors] only had few abnormal degenerates. Homosexuals, called Urnings, were drowned in swamps... That was not a punishment, but simply the termination of such an abnormal life. They had to be removed just like when we pull out nettles, stack them and burn them. It was not a question of revenge but simply that they had to be done away with.

This perception of the bog bodies found throughout much of northern Europe created its own dark strain in Nazi ideology, in which human beings were compared to unwanted plant life, rooted out and destroyed for the good of the whole. Himmler's notions were extended and added to by others, most notably the German bog body specialist Alfred Dieck, whose findings were of great importance to the regime.* According to him, the victims were often those who had 'betrayed the nation', whether in terms of self-mutilation to avoid service to their tribe or having a relationship with someone outside of their community. In this, Dieck's work was seen to be justifying the role of both Himmler and the SS in terms of the history of their country, offering evidence that this was the way the tribes observed by Tacitus had acted before. On one level, it allowed the idealisation of the race to take place. The portrayal of the German Professor Keil, who appears in Michel Tournier's novel *The Erl-King*, may be fictional, but in a scene when a bog body is discovered in 'the peaty eternity' of the moor it illustrates many Nazi attitudes to the German past. He compares the final meal of the individual, a gruel made largely from knotweed, bindweed and daisies, to the last supper taken by Jesus and his disciples, making a connection between this pagan death and that of Christ, which occurred at roughly the same historical period. Examining the head of another figure that is buried nearby, he leaps to the conclusion that this was his wife, 'the ancient Germans being strictly monogamous, as you know'.

Peculiarly, it is this attraction to the purity of the natural world of the German past that both underpinned and allowed their perversion of it, offering those who were believers in Nazism permission to destroy the lives of others, be they gypsies, Jews, homosexuals or those whose political or religious ideas were different to those of the majority of the tribe. This even extended in the Nurnberg Laws in 1935 to proscribing Germans marrying those who did not belong

* Much of his work is now believed to be fraudulent, based on false and fabricated evidence.

to their 'race'. It justified, too, the shaving of women's heads when they stepped across this imaginary line, a punishment for 'fraternising' with the enemy not just within German borders but elsewhere.

Yet even then there were exceptions, occasions when paradoxically the bog might reveal all that is ideal about the German nation, showing the muddle and confusion at the centre of Nazi 'thought'. The mummified bog body of a 14-year-old found in Drobnitz – now part of Poland – was hailed as '*eine germanische Schönheit*' (a German beauty) when she was discovered in 1939.

Sometimes people see what they wish to see…

CHAPTER TWELVE

Schwarztorf (German) – Black Peat

It's strange the way people can be brought to knowledge of their own history.

In my childhood, some was provided in the discomfort of the draughts and hard-backed chairs of the Ness Hall at the edge of my village. We used to drag the latter from where they were stacked beside the building's walls before we settled down to watch that week's edition of Pathé news, waiting patiently until the projectionist Iain Archie MacMillan set up his equipment on stage. He would take dark reels from steel cases, ensuring they were fixed on the wheel of his machine. One of us would dash to switch off the lights before the evening started, watching dark lines worm and wriggle on the screen. A moment or two later and a large rooster would appear, crowing high and proud on a wooden fence. His call would be our signal to start rustling the packets of sweets we had brought from home with us, a constant accompaniment to the soundtrack of the films the Highlands and Islands Film Guild brought to our district – or at least on those occasions when Iain's van wasn't stuck in a snowdrift in the village of Shawbost, an event that once interrupted the supply of movies to our home for some ten weeks or more.

Among the epics that arrived there was *Culloden*, a documentary drama devised by Peter Watkins. Unlike the James Bond and Man from U.N.C.L.E. films with their unknown and exotic backgrounds, this one featured a similar landscape to the one we looked at every day. It featured, too, the battle on Drumossie Moor that brought an end to the chances of the Jacobite dynasty ever gaining the British crown. More importantly, it portrayed the bloodshed and slaughter that took place both during and after the battle when the Highland forces were led into disaster by such officers as the 'frequently intoxicated' Sir John MacDonald, and John William O'Sullivan, an Irishman who, the words provided by the commentator to the film informed us, possessed 'a vanity superseded only by his lack of wisdom'. In authority over them all was the vainglorious figure of Prince Charles Edward Stewart, eager to escape the smoke of battle long before the final swing of a claymore.

And then there was the episode of *Dr Who* I saw during my childhood. Titled 'The Highlanders', it showed the Time Lord in the company of a young clansman called Jamie MacCrimmon, whose tattered, frayed kilt swung and swayed above lucky white heather as he and his companions raced from the approach of the Redcoats. On our black-and-white screen, it looked as though an eternal dusk had fallen on the landscape, late mists shrouding the moor that were disturbed only by the loud and somewhat ungrammatical shouts of 'Sassenachs' littering the script. Together with a few scenes from the novel *Kidnapped*, in which Alan Breck Stewart and David Balfour endure a long and arduous trek across Rannoch Moor, this became part of my fantasy life across the heather. I would imagine Hanoverian heels were hot on my trail as I clambered ridges and slopes, bending down to drink from a flowing stream with the cracked cup of my fingers, believing, in the ominous quality of each silence, that King George's Redcoats were not far away, knowing there was a possible escape in a Tardis in the offing. If I had encountered it at the time, it wouldn't have been difficult to picture myself as Jakob in Ann Michaels' poetic novel *Fugitive Pieces*, leaping into a peat bog to escape my attackers, lying there until my skin took on the shade of dark leather, lines like small cracks appearing on the palms of my hands and soles of my feet.

In one way, my imaginings were not so fanciful. The moor has been a place of conflict for much of the history of these islands – from Glen Shiel to Bodmin Moor, Marston Moor to Hedgely Moor, from long before the Wars of the Roses in the fifteenth century to the Jacobite uprisings of the eighteenth century. The terrain of the moorland provided much of the open ground where troops could meet and either battle with one another or else, as in the Highlands and the west coast of Ireland, offer slopes and crests of hills where one side could wait for the other, a quick and sudden skirmish like the one in West Cork where the Irish politician Michael Collins was shot. It was also a location where deserters from armies concealed themselves, hiding among its desolation from the rigidity of regimental life, whether in Frederick the Great's Prussia or the edges of these islands.

This is true especially of the opposite end of Scotland from my own where in the latter half of the seventeenth century the 'Killing Times' occurred. Drive around the moors of Ayrshire, Dumfriesshire and Galloway and there are a multiplicity of tombs and memorials dedicated to those among the Presbyterian or Covenanting tradition who fought the Royalist forces led by men like James Graham, the Marquess of Montrose and Alasdair MacColla, the Highland chieftain. Walk there, too, and you will find various reminders of the conventicles that also occurred in this moorland, where those opposed to the episcopal form of government imposed upon the church in Scotland by Charles II held their religious meetings.* Often

* In the West Highlands, there is evidence that the Catholic tradition also used the moorland in similar ways, as was done over much of Ireland, the people of that faith heading out to its empty acres to practise their faith and receive mass from the hands of a priest. There are many Mass Rocks in existence, sometimes marked by the sign of the cross or the shape of a chalice. This is sometimes recorded in the place names. For instance, there is both a *Creagan an t-Sagairt* or Priest's Rock and an *Allt na h-Aiffrin* or Mass Burn near the township of Invergarry. The Free Church of Scotland, too, when it was set up in the mid-nineteenth century, also frequently met outdoors, having been deprived of a church building. An example of this are the Worship Stones in Scourie in Sutherland.

surrounded by a circle of stones, such places are in evidence throughout Dumfries and Galloway, like Skeoch Hill near Dumfries, and Moniaive Altry Hill. Indeed, there are almost as many examples of this as there are of bloodshed and brutality in the area. This was so excessive at times that the Scottish man of letters Allan Massie once informed me of the words of an officer in the Covenanters' Army under Sir David Leslie. The soldier was said to have turned to a church minister who was urging on their troops to kill more Irishmen in the aftermath of the Battle of Philiphaugh in 1645:

'Master John, have you no had your fill of blood?'

There were moments in that civil war when it seemed that the officer's question might be stretched beyond meaning, as if what both sides had in mind was to fertilise the barren soil of these moorlands with human blood and carnage. From individual assassinations to wholescale battles, the moors of Dumfriesshire, Galloway and Ayrshire were awash with red, yet some never seemed satisfied with the quantities that had been spilled. In its force and scale – as the Scottish poet from this region, Hugh MacDiarmid, wrote in a short poem about the Covenanters many years later – it is:

> the wind of God,
> Like standing on a mountain-top in a gale
> Binding, compelling …

This kind of opposition to authority happened in similar locations even in near-contemporary times in places like the Baltic States where groups such as 'Friends of the Forest' used the cover of both woodland – as their name implies – and peatland in a partisan war against the Soviet Union, one that went on between 1944 and 1956.* Initially, the men

* This was the most active period of resistance, one that ended after the failure of the West to support the Hungarian Uprising in any meaningful way during that year. Nevertheless, some resistance

and women involved in this struggle used to stay in the farmhouses and homes of those who were sympathetic to their cause. This changed when the NKVD or Soviet government began to gather information and intelligence on the sympathisers. Aware that the inhabitants of these houses were being sent to Siberia for harbouring them, the Estonian rebels and their counterparts in the other Baltic states had to find new homes in order to continue their struggle. They discovered these in the dense thickets, swamps and bogs of their countryside. During the summer months, the endless light was their greatest enemy. Forced to be vigilant, they slept either in the open or within tents, hay barns and sheds used to store peat. In the autumn months, they were forced to look for other forms of shelter. As a result, they built bunkers in various locations, including small hills and ridges in swampland and bogs. Highly camouflaged, these structures were of various sizes and scales, with some containing trenches and embrasures to allow them to glimpse what was happening in the outside world. These bunkers even possessed emergency escape routes in case occupants were ever surprised by Soviet troops. As recorded in Mart Laar's *The Forgotten War*, it was all part of an attempt to prepare: 'an armed rising against the Soviet Union at such a time when England and the United States go to war against the Soviet Union or when a political coup occurs in the Soviet Union itself'.

These 'wars of resistance', however, are now exceptional in terms of conflict rather than customary, as they were in earlier centuries. As can be seen throughout the world, from Basra to Chechnya, fighting in moor and bogland has now given way to streets and bombed-out buildings. For much of the twentieth century, mechanised warfare has been less effective in its former killing fields of the moorland

---

continued until the late 1970s when a number of the most important figures in the Friends of the Forest were shot.

than elsewhere. For instance, during the Winter War between Finland and the Soviet Union, the fact that the former's terrain was mainly moorland and forest proved a major problem for the Red Army. The sheer abundance of this kind of terrain prevented the blitzkrieg that the Soviet Army generals had planned for their smaller, Western neighbours. There were no paved roads in much of the border area and few dirt or gravel roads. Both forest and swampland were largely trackless. As a result, their tanks and trunks became bogged down time and time again, rendering them useless for conflict. Paradoxically, it was when the weather became colder with snow layering the ground that matters began to improve for these mechanised divisions. They could then move quickly over frozen terrain and bodies of water.

One reason for this is that together – thankfully – with this kind of conflict coming to a close, the moor that was the setting for it has also disappeared. Let us take my time-travelling companion of Dr Who, Jamie MacCrimmon. If he had been transported to, say, the Germany of his historical period, the mid-eighteenth century, he would have seen a landscape that was not unlike the one he and the good Doctor had witnessed in the aftermath of Culloden. Together with sand, scrub and great stretches of water, there would have been ridges of uncultivated peatland similar to the landscape Alan Breck Stewart and David Balfour scuttled over. Fast-forward to, say, the 1930s and where once was peat, scrub or sand, he would have witnessed a mass of green fields similar to those I saw from my train window as it raced across much of Germany, taking me from the border with Denmark to Bremen and beyond. Holstein cattle grazed where men had once dug out fuel. Streams, rivers and canals flowed where once there was bog or marsh. That this change had happened was all due to the notion that Nature had to be challenged and defeated in order to provide a better way of life for Man. In this, they followed the battle cry of one of my fellow Scots, James Dunbar, a philosopher who announced: 'Let us wage war on the elements, not with our own kind; to recover, if one may say so, our patrimony from Chaos, and not to add to its empire.'

There was undoubtedly more than an element of chaos within the country we now know as Germany both during and before the mid- to late-eighteenth century. With its assortment of small states and kingdoms, it often provided the battlefield for stronger neighbouring nations. (This is one reason why its unity, once achieved, was so highly prized and, possibly, so menacing to others at one time. The citizens of these states knew all too well the perils of being victims.) Disease and famine were rife. Only one individual in ten reached the age of 60. Half of all children died in their first decade. Much of this was – as I have already noted – the result of the barren nature of a great deal of its terrain. To alter this, there had to be as much weight and emphasis given to the use of a ploughshare as the wielding of a German sword. A variety of battles had to be fought at the same time. One was with its external enemies, whether they were Danes, Poles, French or the people of any other nation. Simultaneously there was also a war to be waged with the earth itself.

There was evidence of this in Papenburg in Emsland not far from the Dutch border. One of the most surreal locations I ever encountered during my travels, the town was a confusing blend of different and distinct localities through which I walked day after day during my time there. A large, impressive shipyard stood at one end of town. It was the last remaining in a community which once sported as many gantries as did the banks of the Clyde in my youth, with 23 dockyards at one time within its boundaries. (In the nineteenth century, ships used to voyage across the Atlantic from the port.) Nearby was a modern shopping centre crammed with glass-fronted stores, each one attempting to beckon customers within its doors. A short distance away was the old centre – a red-brick town hall standing near a hump-backed bridge which seemed to be permanently occupied by an army of cyclists, both local and visitor, wheeling their way across the endless road that stretched through the town. It was this way in which Papenburg reminded me of my native district of Ness. The houses were all set – at most – a mile or two from the main road, its 38,000 or so inhabitants not concentrated in one area but stretched out along its length. If most towns and cities are like a tight, bunched ball of wool, Papenburg is one that has unravelled, spun out in a thin strand that stretched for kilometre upon kilometre.

It was also like my native Ness in part of its history. I walked (and walked and walked) to discover the von Velen museum within the town's borders. It provided a reminder that the community's foundations had been built on peat when it became Germany's first fenland colony in 1630, the township's founder Dietrich von Velen providing the original settler families with a piece of moorland some 50 metres wide and 400 metres long. Without rent to pay for a period of ten years, here was fuel to burn and land to try to cultivate. In this environment, there is little doubt which was the most difficult challenge. The museum marked the progress in housing. The sod hut: home of the first who arrived there, smoke seeping from the mixture of turf, birch trunks and twigs. The Heathkate: with its fire-resistant wall, the animals crammed into its tight and narrow space. The Torfgräberhaus, which, for all that it was constructed from brick, bore some resemblance to the Hebridean blackhouse, with the cattle at the far end of the house. Finally, the Muttschifferhaus, the red-brick building that arrived when prosperity came and there was work to be found at the shipyard.

Yet there were differences. Once again, there were the long days out on the moor where industrial peat cutting was involved, a way of life shared in these islands only by those rare individuals who were employed as full-time peat cutters, earning a pittance during long dry days in the Scottish Highlands and Islands and other areas. On the edge of Germany, the fuel was brought – not by creel on the back of a crouching woman or a Shetland pony – but from the places in which it was cut on a peat ship known as *Prahm*, *Barge* or *Muttschiff*. One of these is found outside the museum. Two figures stand with ropes at the bow and stern. The *Muttschiff* hauls the vessel down the canal by his muscle power. At the stern, the figure they termed 'Jan' also clutched a rope – or *trillkeboom* – which kept the barge still, making sure it stayed centred within the borders of the canal. The barge brought fuel to the factories producing the red bricks that still predominate throughout the landscape of this part of Germany, one that seemed flat and endless under its wide expanse of sky.

Public transport was infrequent. Sometimes it even seemed to operate in disguise. Stand at a bus-stop for a while and a minibus would wheel up, no destination or markings on it. (It can be a little like that in my native Ness.) Sometimes, there would be no transport whatsoever. There was no bus, for instance, between the post office in the village of Esterwegen and the memorial to the Emsland camps which I planned to visit on its outskirts. Only an occasional car, lorry or the clink of a cyclist racing between home and town whirled past, the last of these with a shouted greeting, the wave of a hand. On either side of me were open fields, often maize or sugar beet, broken sometimes by a thin barrier of trees that formed a boundary to the road. Poor soil, that much I knew, and that at a glance, my crofting instincts coming to the fore. Here and there houses stood. Not enough even to bring some shadow or relief from the punishing heat of the sun in that early day of September. I took a swig from the bottle of apple juice I carried in the bag around my shoulder, conscious that if I put it to my lips one more time, I might drain it dry and be left reeling as a result of the unfamiliar warmth of the afternoon. You can take a boy from the islands of Scotland but you can't take the chill of these islands out of the boy. I longed for rain, a lash or two of wind.

My journey hadn't started off like this. When the tourist guide had pointed out to me the road from town, the heat hadn't been quite so relentless, the sun not so powerful. I had obtained the benefit, too, of being in the shade of a local Heimatmuseum with its statue of a peat cutter with a cap pressed on his head, a couple of peat spades and a water-flagon draped over his shoulder. Thinking of my own failure to bring enough liquid with me, I envied that carved figure his foresight, forgetting he was stone and not flesh – the strength of my delusion such that I wondered if the heat of the sun was beginning to affect me.

Yet still I marched, recalling the rhythm and words of an old song I had heard about while researching for this book. Originally titled 'Die Moorsoldaten', it had first been sung at the destination for which I was – eventually, slowly – bound. Over the years it had been translated, renamed 'Peat Bog Soldiers' and sung in English by the likes of Pete Seeger, the Scottish folk group the McCalmans, Luke Kelly and the Dubliners, the great Paul Robeson. Versions of it had echoed across the battlefields of the Spanish Civil War, Communist and Socialist meetings throughout the Continent.

Far and wide as the eye can wander,
Heath and bog are everywhere.
Not a bird sings out to cheer us.
Oaks are standing gaunt and bare.
We are the peat bog soldiers,
Marching with our spades to the moor.
We are the peat bog soldiers,
Marching with our spades to the moor.
Up and down the guards are marching,

> No one, no one can get through.
> Flight would mean a sure death facing,
> Guns and barbed wire block our view.
> We are the peat bog soldiers,
> Marching with our spades to the moor.
> We are the peat bog soldiers,
> Marching with our spades to the moor.

Yet it had first been sung by those who had been imprisoned at my destination by the Nazi government. Esterwegen was one of 15 concentration camps to be found in the districts of both Emsland and nearby Bentheim between 1933 and 1945. Some of these were POW camps. However, the first had been set up in the area by Hermann Göring in his role as minister of the interior for Prussia in Hitler's initial government in 1933. Their purpose was twofold. They were, of course, punitive, but also designed to reclaim the peatland that covered vast areas of this part of their nation. For all that it had been part of their thought processes since the eighteenth century and the time of Friedrich the Great of Prussia, successive German governments began to spend greater time, energy and money on the nation's moors after their country's defeat in 1919. There were reasons for this. They had lost territory at Versailles and sought to use every acre of land within their borders. In the thirties, it also became a way of employing those who were out of work within the larger towns and cities. They could, if they were so inclined, find new homes for themselves on this newly drained land, singing songs that declared 'our spades are weapons of honour, our islands are camps on the moor'. Observing them and the work that was going on, the German writer Hans Plug experienced a rapturous vision, declaring:

> Just the word Emsland has come to stand for the programme of economic and political reconstruction in

> Germany. Living in the wasteland of the moors in remote camps is not easy, but the differences between the cohorts of working men dissolve through work, and one day fruitful fields and meadows will have been created out of unculti vated wasteland.

By the time, footsore and weary, I reached the site of the camp, I was all too aware that there was little evidence of any of Hans Plug's prophecies ever coming true. Fruitful fields and meadows were there none. Only more and more crops of maize and sugar beet. Neither was there any sign of the groups he had watched ever achieving redemption through their labours. Not unless salvation and deliverance were measured by the high concrete posts and barbed wire that marked its former boundaries. Within the grounds of the museum, there were ochre-shaded sculptures and entrances to commemorate the prisoners that had once been there. A bridge of the same colour spanned the area from the camp to where they had worked each day, digging and draining the moor. Clusters of green trees took root in the spaces where the camp buildings had once stood, their branches stirring back and forth. All this was to remind people of how men had once worked there, setting off with their tools on their shoulders, singing both the authorised songs that praised the Führer and called upon them to make themselves 'ready at every hour to serve Germany with all their thoughts' and that which in its full version read as follows, the words – composed by the prisoner Wolfgang Langhoff among others – even repeated by the SS guards who saw themselves in some twisted fashion also as the Peat Bog Soldiers of the title.

Die Moorsoldaten

Wohin auch das Auge blicket.
Moor und Heide nur ringsum.
Vogelsang uns nicht erquicket.
Eichen stehen kahl und krumm.

Wir sind die Moorsoldaten und
ziehen mit den Spaten ins Moor.
Wir sind die Moorsoldaten und
ziehen mit den Spaten ins Moor.

Hier in dieser öden Heide
ist das Lager aufgebaut,
wo wir fern von jeder Freude
hinter Stacheldraht verstaut.

Wir sind die Moorsoldaten usw

Morgens ziehen die Kolonnen
in das Moor zur Arbeit hin.
Graben bei dem Brand der Sonne,
doch zur Heimat steht der Sinn.

Wir sind die Moorsoldaten usw

Heimwärts, heimwärts jeder sehnet,
zu den Eltern, Weib und Kind.
Manche Brust ein Seufzer dehnet,
weil wir hier gefangen sind.

Wir sind die Moorsoldaten usw

Peat Bog Soldiers (literal translation)

Wherever the eye gazes,
Bog and heath all around.
No chirping of birds entertains us.
Oaks are standing bare and crooked.

We are the peat bog soldiers
And we are marching with our spade into the bog.
We are the peat bog soldiers
And we are marching with our spade into the bog.

Here inside this barren marshland,
the camp is built up,
Where we are, far from any joy,
stowed away behind barbed wire.

We are the peat bog soldiers etc.

In the morning, the columns
march towards the moor to work,
digging under the searing sun
but our mind yearns toward our homeland.

We are the peat bog soldiers etc.

Homeward, homeward
everyone yearns to the parents, wife and child.
Some chests are widened by a sigh
because we are caught in here.

We are the peat bog soldiers etc.

Auf und nieder geh'n die Posten, *Keiner, keiner, kann hindurch.* *Flucht wird nur das Leben kosten,* *vierfach ist umzäunt die Burg.*	Up and down the guards are walking. Nobody, nobody can get through. Escape would only cost the life. Four fences secure the castle.
Wir sind die Moorsoldaten usw	We are the peat bog soldiers etc.
Doch für uns gibt es kein Klagen, *ewig kann nicht Winter sein,* *einmal werden froh wir sagen:* *Heimat du bist wieder mein.*	But for us there is no clamouring. it can't be an endless winter. one day we'll say happily: 'Homeland, you are mine again!'
Dann zieh'n die Moorsoldaten *nicht mehr mit den Spaten ins Moor.*	Then will the peat bog soldiers march no more with the spades to the bog.
Dann zieh'n die Moorsoldaten *nicht mehr mit den Spaten ins Moor.*	Then will the peat bog soldiers march no more with the spades to the bog.

I read about these prisoners in the steel shell of the central museum. Many of its documents and photographs were on display behind chicken wire and glass, within the cold and – rightly – inhuman building. I read and saw, too, of how the prisoners had on their first day there looked out on both barracks and watchtowers, the swastika flapping on a flagpole. Other than that, it was, Langhoff wrote, 'endless moorland, as far as one can see. Brown and black and ditches running through.' For all the bleakness of that landscape, one emerged with an even more desolate view of the human spirit from all that was included in the exhibition. There were accounts of the many atrocities inflicted on those who had been imprisoned in the camp. They all provided testimony of how groups like the SA and the SS at a later stage behaved towards those who were under their control. They, too, gave dark proof of the underlying seriousness of the pun employed by Himmler when he spoke about those whom he threatened with a period in Emsland.

'You wait, I'll teach you *mores* [manners]. I'm sending you into the moor.'

In the early years of the Emsland camps, some of the prisoners were Jewish; a number killed within days of their arrival. One who survived for a time was the SPD politician Ernst Heilmann, who was imprisoned in both the Börgermoor and Esterwegen camps. His welcome to the former included being flung into a hole, together with four other Jewish prisoners. Among them was the Rabbi Max Abraham, who was told that the dung-pit into which he had been thrown was a place where he could hold 'divine service'. Continual humiliations followed, particularly for Heilmann who was regularly beaten and smeared with excrement. Ernst attempted to commit suicide by trying to escape. The guards, instead, shot him in the thigh. In the end, it was the local population who brought an end to the excesses of the SS members who controlled the camp. They reported their concerns to the Prussian Ministry of the Interior. The Jewish prisoners and some of the more prominent politicians were sent to Lichtenburg, near Wittenberg, where, for all the undoubted brutality, there was an end to the 'special treatment'. Sadly, this did not save Heilmann, who was murdered in Buchenwald in 1940.

The majority of prisoners in Esterwegen, however, had been left-wing intellectuals, Communists and trade unionists who had fallen out with the regime; many were members of the *Nacht und Nebel* (Night and Fog) movement which had been the most vehement of those hostile to Hitler and his supporters. One famous individual was Werner Finck, a comedian who allowed his audience to complete the jokes he made against the Nazi entourage, slowing down his delivery to allow the members of the Waffen SS present to take notes of all that was said. Another was Karl Germer, who was a friend, associate and later the successor of the 'wickedest man in the world', the occultist and ritual magician, Aleister Crowley. A number, too, were prominent political figures, such as Julius Leber, a friend of Willy Brandt, who was eventually tried and sentenced for his alleged role in the attempted assassination of Hitler in 1944. Another was a German pacifist called Carl von Ossietzky, who had been convicted of high treason by the government in 1931 for revealing how Nazi Germany was secretly rearming itself. In 1936, he was awarded the Nobel Peace Prize for this action, an honour he was not permitted to travel to Stockholm and receive. The account of the Red Cross representative who was sent to inspect the conditions under which he was held is revealing. He reported that 'the SS came back with a shivering man, pale as death, a poor creature who was unable to feel anything. All his teeth were broken and he had a broken leg. I came to him for a handshake. He did not respond.' The report concluded that he was 'a human being who had reached the utter limits of what could be borne'.

Conscious that the savagery of this treatment had become a political problem for the Nazi regime, the regime sent him to a civil hospital, where he died in 1938.

The camps in Emsland are not the only examples of how the Nazi party used – what they considered – wasteland as sites for some of their camps. Dachau, for all its grotesque reputation today, was once an artists' colony, inspiring painters like Christian Morgenstern, Carl Spitzweg, Eduard Schleich the Elder or Adolf Hölzel, Ludwig Dill and Arthur Langhammer to include that landscape in their work. (The town's website boasts that at the end of the nineteenth century, and beginning of the twentieth century, 'every tenth person in Dachau was an artist'.) Auschwitz straddles a reclaimed marsh. Belsen stands in Lüneburg Heath – the stretch of moorland where German forces surrendered at the end of the Second World

War. It is tempting to conclude that this links with their attitude to the Slavs, Poles and others who were held prisoner in these camps. They regarded those people as passive and feckless, incapable – unlike, for instance, both the Germans and Dutch – of transforming through industry and digging the landscape they had inherited.* Their perceived backwardness was evident in their inability to drain either moor or marsh, utilising their environment for their own profit and benefit. As David Blackbourn points out in his wonderful book *The Conquest of Nature*: 'When Primo Levi later called Auschwitz the ultimate draining point of the German universe, he had thought himself into the heads of his persecutors, for whom drainage was both metaphor and reality.'

Yet all this, too, existed alongside a nostalgia for the moorland that is present in many German writers, such as Theodor Fontane and Theodor Storm, the latter often idealising his native heath in his work.† It can be seen in the signs I noticed while driving round the county of Sutherland, where the sheer frequency of drivers from Germany warranted the warnings in their own language about the presence of lambs found by the roadsides. Most particularly, it can be found in locations like Lüneburg Heath, which I

* There is a similar attitude to be found in C. S. Andrews' *A Man of No Property*, where he criticises his own nation for failing to do anything constructive with the Irish moorland.

† Even Hitler succumbed to this on occasion. In February 1941, he issued a moor edict, calling on the existing moorlands to be preserved. The reason he cited for this was *Versteppung* or 'desertification', the effect that, similar to what happened in the Netherlands, drainage was having on the water table, contributing to flooding in various locations. Some argued against this, making the case that peat was an essential element for Germany in their war against the Allies. A short time later and the edict was quietly withdrawn.

visited on my travels through Germany. A road looped around part of it, one which I journeyed across with a busload of others, mostly older than me. Together we snatched looks at the countryside outside the window. There were flurries of purple on display, clearly heather flushed and bright and at its best in the month of August, shadowed by an occasional juniper tree or two. Sheep grazed. Occasionally, in townships that seemed little more than clusters of bars and restaurants, we watched people sitting at tables nursing giant glasses of beer or gorging on equally gargantuan meals. Sometimes, too, the passengers left their seats to go for a stroll across the heather or, more specifically, the path that trailed across the heath. A couple kitted out in waterproof windcheaters. A man dressed like a caricature of a German hiker. Shorts. Knee-high socks. A collarless jacket. A hat decorated with an array of bird feathers. Part of my mind couldn't help but wonder if he had pointed a rifle in their direction, claiming their plumage as a prize.

I never joined them. There was a mixture of reasons for this. A sleepless night after travelling from Berlin and arriving in the town of Lüneburg the night before. A slight feeling of being out of sorts. But most of all there was my sense that, despite the plumes of heather, this heath was not my idea of moorland. With its paths and hordes of visitors, it seemed far too tame, timid and denatured to be that. There were no signs of the clouded crests and tumbledown burns of the island of Harris, where I had sometimes walked or scrambled with my son to places like Sròn Uladal with its overhanging cliff of grey rock. Neither could I imagine the whirl of the *guilbneach* (curlew) as it wept and cried at our arrival when we walked along the track towards the east side of Benbecula, the dampness of the land squelching and sucking at our boots. It did not have the magnificence, either, of Eaval on North Uist, where I could peer down at a moorland that looked a little like a lace handkerchief, water glistening in all the tiny cuts and tears that formed a rich and changing pattern on the surface of the land. These places were bog. These were heather. These were genuine, authentic moors.

Yet all this was to overlook that these acres too had stories, some which retired headteacher John McMillan told me when I visited his cottage in Lochcarron. A slight, impish individual with a talent for adventure, storytelling and clinking spoons rhythmically together upon his upper thigh, he has retired to the edge of the moorland, the windows of his home permitting on a rare, clear day* glimpses of a cluster of islands, varying in size and shape from the tiny Inner Hebridean isle of Scalpay to the heights of Raasay, which the Gaelic poet Sorley MacLean immortalised in his work and where Boswell enjoyed a Highland dance on the crest of Dùn Cana. Behind them, stretched like the wings of a bird – this similarity the basis of its Gaelic name – was the island of Skye. Opposite was the village of Ardmore, about which Norman MacCaig had written:

> The sea, the sun
> Are the next stage, with nothing in between.
> A quick place this to know your journey done.

For John McMillan, there is no sense yet of his journey being done. Over the last few years, this man, now in his mid-seventies, has travelled far and wide, diving in the Philippines, travelling round the islands of the South Pacific,

* The Skye and Lochalsh area – of which Lochcarron is a part – suffers from roughly the same annual rainfall as the Amazon rainforest. There are, however, significant differences, as John pointed out. There is no dry season in the West Highlands of Scotland.

often astonishing the natives by wearing his kilt – a form of clothing he sometimes adopted in colder, more inclement weather. By his fireside, though, he told me of his younger days when, between the years 1967 and 1972, he had been employed as a maths teacher in Gloucester School in Hohne, Germany. It was a place, John discovered, that possessed as many layers of history as the peatland on which it had been built. Behind his house, hidden by a clump of trees, was a railway siding where the prisoners bound for the horrors of Belsen concentration camp were delivered and held for a short time before being marched to its gates. Nearby, too, there are mass graves: those of the inhabitants of the camps whom Allied troops had arrived too late to rescue, and those of thousands of soldiers, with a total of 50,000 Russians being buried in one location. Often these burial grounds were marked with heather. Sometimes, too, the roots of trees turned and twisted among the corpses, bone indistinguishable from wood. His house and school stood not far away from evidence of another conflict. The Iron Curtain between countries of the Eastern Bloc, including the former DDR or East Germany, was only a few miles down the road. The airfield at Fassberg, a short distance to the northeast, had played an important role in the Berlin Airlift in 1948–49, with flights bringing food and other supplies to that beleaguered city. Some of the planes even took on the role of coal merchants, 'The Last Ton of Coal from Fassberg' emblazoned in black on their side when they delivered fuel to Berlin.

Yet the moorland also performed a role to which it has become accustomed over the last few years – training the armed forces for the possibility of war. While teaching in Stornoway, I can recall some of my pupils, red-faced and breathless, arriving in my room. 'Sir! Sir!' they announced. 'I was arrested by the troops yesterday. They pointed their guns at me!' Apparently, while heading out on a fishing expedition, the boys had blundered into a NATO exercise and been mistaken for enemy soldiers. I can recall, too, how the armed forces used the Isle of Harris, that stretch of moor and rock Kubrick filmed to imitate the surface of the moon in the film *2001: A Space Odyssey*, as a way of preparing their men for the Falklands War. The same was true of Emsland, where men like Carl von Ossietzky had once been to set to work to break the damp and choking hold of peat upon vast acres. The moors, too, were often the locations for troops and military exercises, where men sought to counter the menace of the Soviet threat against their borders.

Given his temperament, it wasn't long before John showed me the light side of even this situation. Over a glass or two of whisky, he spoke of how a convoy of military vehicles towing a trailer loaded with Honest John missiles, some 8 metres (27 feet) long, had arrived in a cul-de-sac near the firing range not far from Hohne, his home during his time in Germany. There was no way back that did not involve crossing pavements or reversing over neatly manicured lawns. Disturbed from his reading by the ensuing uproar, John was amused by one of the regimental commanding officers who stayed in one of the houses, bellowing at a young captain. His face was red and furious, his words spitting like bullets from his lips.

'Get that missile off my lawn! Did they never teach you to "read" a bloody map at Sandhurst, you idiot? Did you never learn to read at all, you incompetent buffoon? There is a huge road sign about six feet square at the entrance to this road saying, ACCESS TO OFFICERS' MARRIED QUARTERS ONLY. CAUTION – CHILDREN PLAYING. And you drove past the bloody thing! Did you not see them?'

'And all the time, the captain stood trembling before this onslaught,' John said, imitating the man's performance, 'his arm going up and down like a semaphore signal, saluting and apologising like someone from a Monty

Python sketch. "Frightfully sorry. Sorry, sir. Yes, sir. No, sir. I mean, yes, sir.'"

Perhaps these are words that should be uttered by a representative of mankind, saying sorry for the way we have wielded weapons for generations across the war-torn expanses of the world's moors.

PART FIVE

Rathad an Isein – The Birds' Path

CURLEW

The evenings that our village football matches
rolled on too long, she was our alarm,
watching out for close of day, soaring up to warn
us night was coming in and we'd soon fail to see
either ball or goal. She'd be there to scold us,
become linesman and referee
with neither flag nor whistle.
Instead, that rippling note
flapping free from wing or throat,
informing us we were in injury time,
just like her absence now on moor and shore
offers the same message to members of mankind.

CHAPTER THIRTEEN

Sùil-chruthaich (Scottish Gaelic) – Quagmire or Bog, literally 'the eye of creation'

Of all the people I met on my wanderings, there was no doubt that the Dutchman Jans de Vries was the most remarkable.

A strong, broad-shouldered individual, he looked somewhat incongruous as he sat at his desk in the office of the Bargerveen Nature Reserve. A green cap with the park's insignia was pressed on his head. He wore, too, a cargo vest of the same shade with an array of zips and pockets. A mobile phone was clearly packed away inside one, the glint of a variety of pens peeking from another. Binoculars hung around his neck. When Roel, Aleid and I arrived to see him, he stretched out a freckled hand to greet us, not rising from his chair. It was only when we went to leave with him a short time later that I worked out the reason for this. Jans suffered from a degenerative illness that had left his legs paralysed and immobilised.

Not that this stopped him. A short while later and he was out on his wheelchair escorting us around the reserve. His journey was made possible by a firm called Ziesel, the all-terrain vehicle on which he travelled having the formidable framework of a tank. Its bulk and tracks enabled Jans to travel across the 2,100 acres of bogland that formed much of the reserve. With a simple movement of his hand, he could go forward and back over its expanse, relaxed and talking about the splendour of his vehicle – 'There's a model with a snow-plough ... There's one with mirrors and so on for the public road.' – and also about the moor that's all around us, pointing out everything in its midst.

He indicates the presence of a sundew beside a track, one that blooms upon the moorland 'unseen by men', a plant that – according to an anonymous poem written in the nineteenth century – was believed at one time to 'drink the pure water of the skies'. Instead, of course, the truth is more harsh and uncomfortable. Nature red in frond and leaf, its crimson foliage bristles with tiny tendrils. At the tip of each one, there is a bright droplet of nectar, a feast for the flies drawn to it. Some 20 minutes later and the insects are already being digested, their flight trapped and brought to a halt by the tendrils closing down upon them.

He points out, too, the bright orange tint of bog asphodel a short distance away. It is familiar from my own native territory, and he listens when I tell him that, in Shetland, its shade was used at one time to colour wool as a substitute for the more expensive saffron, part of the extraordinary range of shades that man and woman has plucked from the moor for their own use. He notes the presence of a common blue butterfly flitting through the heather, food for the red-backed shrike that sometimes visits both here and my home in Shetland, seeking sustenance. Some of the diet of both this and other birds flying around Bargerveen is provided by the last remaining house in the estate, which belonged one time to a turf cutter and has been allowed to crumble, the insects scrambling within its walls providing food for the birds that are now its chief residents. He also informs us of the dams that have been built to raise the water level, how sheep and cattle have been used as part of the management of the estate, keeping rush and grass short where this is required. In itself, this makes Bargerveen different from other reserves I have visited elsewhere, where sometimes the abiding impression is the absence of man and domestic animals.

He takes us to the Hill of Truth, a crest of moor that rises a few metres above the remainder of the reserve. It reminds people of the original height of the Bargerveen moorland before peat was cut here, an activity that only stopped in the early nineties, providing work for the members of what was undoubtedly a poor, Catholic community in a largely Protestant country. Its lack of wealth is in evidence even in the tiny dimensions of the local church, one that stands beside a small former convent that now serves as a guest house. Roel, Aleid and I stayed there that night, the remnants of its former purpose, portraits of the Virgin, a crucifix and votive candle or two, surrounding us as we slept. Portraits, too, of various nuns of the order that served here looked down benignly at us when we ate.

This memory of poverty has left its own legacy. I was reminded of this when Jans's mobile phone rang during our visit. There was a short burst of conversation. Apparently, a neighbouring farmer had taken it upon himself to bring a slurry-tanker down the winding roads to his acres, pouring a heady mix of manure and water onto one of his largest fields. There was, clearly, a problem with this. This particular stretch of land had been due to become part of the 2,100 acres of Bargerveen's nature reserve the following year. By making the soil richer and more fertile, the arrival of the contents of that vehicle would delay the process of transforming it into moorland by a few years.

'I do not understand it,' Jans declared.

Yet part of me – and I suspect part of Jans, a native to that area – could. I recall the first time I heard a conversation about 'rewilding' the land, restoring the earth to its natural, uncultivated state. This individual was speaking about how within the borders of her croft she had blocked the drains, allowing the ground to become wet and waterlogged. The fields, too, which her predecessors had tilled and ploughed for oats, potatoes, turnips, were left undisturbed.

I bristled silently about this. It seemed an insult to my ancestors, men like my great-grandfather Stuffan who hoisted and heaved rocks to make the slow movement of a plough possible across the length of a thin strip of land, who had in similar way to Sisyphus stood 'pechan an sweitan, disjaskit, forfeuchan and broun'd aff' as he and his sons had tried to dig drains and enable the growing of crops in this poor soil. This had been their attempt to create their own 'Pleasantvilles, Fair Havens, Murray's Harbours, Heart's Desires' in barren and unpromising territory. And what was more, it was those in authority over them, the equivalents of Leverhulme, the Duke of Sutherland, General van den Bosch, who had encouraged them to act this way in the first instance. Now they were – in the shape of George Monbiot, the environmentalist movement, Greenpeace, Friends of the Earth – demanding that they do the opposite. One could only be flummoxed, flabbergasted and fair affronted at their nerve.

It is in Ireland that this attitude seems at its most intense. Click on to the Barroughter and Clonmoylan Bogs Action Group Facebook page, for instance, and you will see evidence of the fury of a group of individuals in that part of County Galway. It displays a plethora of placards among its postings, as many as Hector Macdonald in the *West Highland Free Press* once dreamed about being brandished by protesters in my native district of Ness – all declaring 'Hands Off Our Bogs', 'Cut Your Turf and Have a Warm Hearth', 'Turf Cutters Are Not Criminals', and even the somewhat ironic 'Protect Your Cultural Heritage'. They even include the masked figure of Donald Trump, shaking the peat-cutting blade that his maternal ancestors in the Isle of Lewis once deployed and demanding an end to government and EU interference with their traditional fuel. Not for nothing did the Irish poet Louis MacNeice describe the 'coal-black turf-banks' of such counties as Mayo in his sequence of poems *The Closing Album* as being 'like the tombs of nameless kings'.

This apparent militancy is similar in many ways to the attitudes that marked the people of my own *Eilean Fraoich* ('Heather Isle'), as in the Gaelic song of the same name that was the anthem of my native isle. The resolution not to be told by others what to do about matters relating to their home turf. The desire to cling on to tradition. That sense of being under siege by the rest of the world. In Ireland's case, this is intensified by the economic uncertainties that far too often have plagued and troubled its people, their distrust of any authority deepened by incidents like the landslides that afflicted Derrybrien some 30 miles away from Barryoughter and Clonmoylan. And there is, too, the fact that peat permeates the life of the Irish, especially in its centre and near its western coastline, in a way that it does not affect the existence of the people of other nations such as the Scots ensconced in the Central Belt and largely indifferent to the moors that surround them. This lack of concern is – to some extent – shared by the English, Dutch, Germans and others to whom that fuel has long been consigned to history. Only the Finns who also use it for their power stations live in such proximity to the presence of the peatland and the moor.

It is this that makes the work of bodies like the Bog of Allen Nature Reserve in Ireland difficult when compared to similar organisations in other nations. It is noticeable, for instance, that the awareness that the peat bogs of Ireland were in peril came initially from outside the country's own borders. A Dutch research student, Matthijs Schouten, visited the country in 1978 and saw in the work of Bord na Móna and other similar bodies evidence that history was repeating itself, that the plundering of peat that had been to the detriment of the Netherlands, in terms of flooding and effects on wildlife, was also occurring there. When he went home, he set up the Stichting Tot Behoud Van de Ierse Venen (the Dutch Foundation for Conservation of Irish Bogs) to obtain funds to purchase three peatlands in Ireland, initially ones in Westmeath, County Galway and Kerry. Other areas have been bought since, the awareness that these peatlands are threatened by man's actions increasingly seeping into the consciousness of the Irish thanks to the hard work and energy of Dr Catherine O'Connell of the Irish Peatland Conservation Council and her team at the Bog of Allen Nature Reserve. Yet it is still the case that, according to their own website, in that country, 'people still think that it is acceptable to cut turf from sites they know have been designated for conservation'.

At this point, it is tempting to mount my own individual 'Hill of Truth' and preach the gospel of peat conservation. I could, for instance, mention the issue of global warming and note that, in a very different way from our ancestors, we all depend on peat. We require it to prevent flooding, especially near low-lying river banks where, all too often, modern housing is being built. We need it as a carbon sink to suck and store away the carbon dioxide that man has produced with the effects of the technology, even the industrial farming, he has created over the last few hundred years. The oceans, forests and moors help us to do this, reducing the release of greenhouse gases which heat up the planet, with 2016 and 2017, for instance, being two of the hottest years on record. As recently reported in the *Guardian*, scientists have declared that the world was last as warm as this around 150,000 years ago, and that it has not experienced such a high level of carbon dioxide for 4 million years. They would also point out that peat is by far the most effective way of dealing with this issue. Trees only trap carbon dioxide for their lifetime. Peat, however, does this for thousands of years, confining twice as much as every forest in the globe.

And some of the truths uttered by scientists are self-evident. Take a walk across moorland and its ability to retain water is obvious. The squelch of boots crossing its breadth is a simple test of that. Scoop up water in your hands on the edge of a peat bog as my great-uncle Alex Smith taught me over 50 years ago and you can see how its depths clean and filter, making that liquid on which we all depend transparent and pure again. Is it really so fanciful – given the fact that peatland exists from the Arctic Circle to New Zealand – to believe that it performs the same function for this planet of ours, ensuring it is fresh and wholesome, healthy and a source of vigour and life? Breathe in the fog of a moor fire, as I did the summer it clouded and choked the town of Stornoway for days, and it is all too easy to accept the idea that it consists of 50 per cent carbon dioxide when it burns.

There is also the contrary notion. Scoop peat upwards with a spade and spill it into the edge of a loch or lake and watch the dark cloud that looms for a time in the water. Contemplate that cloud for a while and you see the problems that the 70 per cent of peat that is damaged or degraded in Scotland can cause.* It can create a host of problems, even silt which – as the landslide in Derrybrien in 2003 showed – can poison the fish in our rivers, affecting, too, their breeding, even the salmon as it struggles to make its way upstream. Silt also costs the organisations and bodies that provide us with water a great deal of expense and money in their efforts to make it unsoiled and spotless again.

It is awareness of these issues that motivates Andrew McBride, who is a Peatland Action Manager in Scotland. He outlines the steps he is undertaking at the moment in his current role. 'Twelve kilometres of peat hags are being reprofiled and repaired just now, to cover the bare and exposed peat with heather, sphagnum and extra layers of peat, in order to trap more carbon. There will be over 10,000 hectares done this year; 20,000 each year until 2032. One of the greatest difficulties is that we have to train digger drivers in all of this. They're used to scooping out drains. In some ways, this is the opposite. We're filling them in.'

Declaring that 'even if the science about global warming turns out to be wrong', he says, 'the changes promoted by modern peatland management still make sense, improving the peatland in all sorts of ways.'

He goes on to list how this peatland management will assist water flow, flood defences, even salmon fishing. 'It'll

* The figures are even higher for the raised bogs found in the Central Belt of Scotland, where 90 per cent is said to be damaged and degraded.

help even lambs and the grouse chicks here in Scotland, preventing them from falling into gaps and pools of water that have opened up on the moor, sometimes stopping them from drowning.'

Another benefit that Andrew foresees is to the cohesion of the community. 'You have to be aware,' he says, 'of the importance of employing local people in key roles in this development. They are the ones who are best equipped to be aware of the web that binds people together in these places, the compromises that have to be made between one person's interests and the next. After all, one man's drainage can sometimes be another's flood. The community has to be given the power to resolve these conflicts. Not outside bodies. That in itself is a big cultural shift from what has happened before.'

I nod, aware of the occasional placard and poster I have seen while travelling across the Highlands and Islands over recent years. All brandished near the homes of local people, their targets are many and varied – from Scottish National Heritage to the Scottish Society for the Protection of Birds, wind farm projects that are planned for glens and the slopes of a Highland mountain. Though there are differences in their messages, they are similar in their distrust and dislike of both science and expertise, symptoms of an age when authority is often distrusted and – sometimes – irrationally questioned. They feel they are being dictated to from on high. And so, as a result of this, the sceptic will shake his head and announce, as one did recently on YouTube, below a video about the preservation of peatland: 'It's amazing how far scientists will go into the realm of speculation to gain their peers' attention', his allies going on to indulge in that modern sport of climate-change denial.

One can understand how we have arrived in this situation, especially in areas like the Scottish Highlands and Islands. A few decades ago, its people were being told to 'improve' their land, with some of my neighbours, for instance, spending a great deal of time, energy and money reseeding peatland, transforming the most unpromising acres into well-drained grassland, rich and green and useful for animal grazing – with the sheep even being granted the courtesy of a shelter belt of trees. We were informed, too, we should use the resources provided by the land, cutting the peat provided by these empty acres rather than purchasing coal or some other form of household heating. Now my fellow islanders are being told the opposite – that it is moral and right to block the drains that our ancestors dug, that we should 'rewild' the ground we once 'reseeded', transforming it back into the bogland it once was, that we should – in some circumstances – refrain from peat cutting. Little wonder then if ordinary people come to believe that only one truth has emerged unchanged from all of this – that the moorland has become a setting for yet another morality tale, albeit one unlike those that General van den Bosch, Heinrich Himmler and even the storytellers of my childhood told in the past.

It does not help either when the weight of the law is imposed with a heavy hand to reinforce its meaning. In the 20th March 2017 edition of the *Irish Independent*, for instance, a young turf cutter was threatened with the possibility of a 500,000-euro fine and three months in prison for cutting peat in a 'designated area of conservation': Moanveanlagh Bog near Listowel in County Kerry. (A few years ago, in August 2013, peat cutting made an unlikely return to being a spectator sport, when 150 turned up to watch a number engaged in working illegally on the turf.) Repressive measures like this may, in fact, turn out in the long run to be self-defeating and counter-productive. It is not, after all, that many generations ago when Pat 'the Brute' Healy and his fellow champion turf cutters were heroes to the Irish – as their equivalents were, too, to the Dutch, Danes, Germans and Scots.

Apart from the scientific and – even worse – the punitive approach, there are other ways in which hearts and minds can be won. Dr Catherine O'Connell, at the Irish Peatland Conservation Council, pinpointed the moment when she became convinced of the value of moorland. The awareness occurred at the age of five, when her parents took her on a caravan holiday to Connemara in the west of Ireland. She recalled: 'During the trip my father took off to the bog one day in search of sods of turf. We had a lovely antique copper and brass fuel container beside our fire in the city, and the turf was to be put there. Never to be burned but to be used as a decoration because it looked nice. So that's why we need to save them, because they are beautiful and every bit of them is precious.'

There are others who regard the moor as a place of healing – not just in the form of the plants that grow there, often regarded as cures for the many illnesses that can ail mankind, but also in the sense of itself. I think of my great-uncle Allan, half-paralysed from his teenage years, spending time exploring its wilderness far from the eyes of others who might seek to judge and pity him. I think, too, of my friend and fellow writer Raymond Soltysek, who grew up in the tenements of Barrhead outside Glasgow:

> with dad who worked away from home and a mum run ragged by five weans, for whom the hills were never an option. Yes, we played cowboys and Indians and went stickleback-fishing in the reservoirs up the Gleniffer Braes, but these were short fair-weather outings. Basically, I never considered myself anything other than occasionally outdoorsy. Then three years ago, a crippling bout of depression drove me to the mountains to find some quiet space for my head, and it's almost an obsession now. There is nothing more emotionally calming than Scotland in October, when settled weather combines with a pellucid, golden light to create a sense of otherworldly peace. I've never had an addictive personality, but being at the summit of Sgoir

Gaibhre, looking down the length of Loch Ossian and the Nevis Range and the Mamores, utterly alone in a landscape the colour of beaten metals, is almost a narcotic experience. It is my drug of choice.

Others like Margaret MacIsaac see it is one of many sources of artistry in her life in North Uist in the Western Isles, describing how she managed even to create paper from peat as part of an art project, lifting a sod of turf from the moor to do this. 'I mixed it all with water in a tub to form a fibrous, viscous soup, and sieved thin layers with a mesh frame, placed on individual J-cloths and left to dry. It then peels off the J-cloth as a sheet of paper. The story of the landscape contained within the page, rather than written upon it.' For some on the continent, most notably in Germany, it is the means by which physical beauty can be obtained. Step into a peat bath, and skin and complexion can be scoured and cleansed.

For me, the love of the moor came from a variety of different sources. There was the pleasure I gained from the birds on the moor that edged the village – the lapwing with its strange quiff, floating and tumbling through the air, lacking gravity in both its demeanour and flight; the curlew with its whirring in the twilight, an eerie final whistle to our football matches out at Sabhal na Caorach,* the sheep's byre,

* My friend Kenny Macleod from the village has suggested that this is a word that has lost its original meaning, that at one time this patch of bare land was called *Sùil na Caorach* or 'the eye of the sheep', and was where several of the crofters' sheep drowned when they returned to the village after grazing further out on the moor. Another friend, Norman Graham, put forward the notion that the peat here was used to provide insulation for the space between the inner and outer walls of the blackhouse, rather than use soil from croftland – or indeed the better quality peat further out on the moor.

possibly the only football pitch with a surface of bare and dusty peat. There was even the menacing line of gulls which once nested a short distance out from the houses in the days before the rubbish lorry clanked and clattered through our community. (All gone now, cleared from their stance by the arrival of the council lidded wheelie bin. People no longer deposit their food waste – or anything else – on a manure heap behind their home.) I have seen, too, that I was not unique in this response. In the past few years, with the help and assistance of the writers' centre at nearby Moniack Mhor, I have been in the company of the remarkable Suzann Barr, who works at Abriachan Forest Trust. During a short, miraculous few hours, she restored my childhood sense of wonder at the moorland, coaxing me once more to enjoy what I loved doing many decades ago, standing alongside other children in a burn that runs through these grounds, scooping mayfly nymphs and pond-skaters from the water, stirring the dirt and grit that lies below with wellington boots. She would stand still and cock an ear, too, listening out for the sounds of birds that flap and fly around the Forest School they have built there. The screech of a peacock. The lilt of a cuckoo. Eyes might rise, too, at the sight of a hawk or a buzzard circling, seeking prey, a small bird that would soon be at the heart of a blizzard of feathers and down. I have enjoyed enough conversations over the years – with, say, the likes of Katie Geraghty at the Bog of Allen Nature Reserve – to know that these childhood experiences are not unique, that they are a vital component in allowing young people to enjoy and appreciate an environment that has too often in the past been feared and disliked.

There was, too, that fateful birthday during my last years in school when my then girlfriend gave me two copies of *Corgi Modern Poets*, bought from Loch Erisort Bookshop in Stornoway. In the beginning, I was reluctant to read them, believing poetry was soft and daft and barely worth considering, but, eventually, the tug of romance defeated the pull of teenage inhibition. I opened the pages of these books and discovered the work of Seamus Heaney, Iain Crichton Smith and the Orkney poet George Mackay Brown. In their words, I discovered there was much that was extraordinary in a world that I considered all too dull and mundane – from peat cutting to walks upon the moorland, from the birds that swept and soared around me to the men and women with whom I shared my village life.

And so I address their ghosts: they have much to answer for.

Charity Tractor Run, Ness, Isle of Lewis
April 2015
To Dòmhnall Blondie, fellow Nessman and postman in Shetland

There are tailbacks all the way from Dell to Port,
congestion in a crofting district, snarl-ups, traffic jams
and tangles as tractors puff and cough to show support
for one whose days might be slammed
short by illness. Men come from throughout the island,
all these ghosts clearing smoke from exhausts,
giving it full throttle. They fix their rear-view mirrors, scan
those behind them that they've already lost.
Aonghas Mòr steaming down the Aird Road to collect
his pension. Doilidh taking home peats.
Dad heading out in winter to correct
some fault in the engine, his feet
leaving prints in mud or snow. And then there is that train
travelling out the peat road, seeking to haul
and take the strain off some tractor stuck in that terrain,
tyres skidding through its surface as it halts and stalls
deep within that morass, sinking further down
till it's engulfed by darkness like these souls
who never – these days – wheel their way around ...

What would the world be, once bereft
Of wet and of wildness? Let them be left,
O let them be left, wildness and wet;
Long live the weeds and the wilderness yet.

'Inversnaid' (extract), Gerard Manley Hopkins

Select Bibliography

C. S. Andrews, *Man of No Property* (Mercier Press, 1982)
William Atkins, *The Moor: Lives, Landscape, Literature* (Faber and Faber, 2014)
Samuel Beckett, *Mercier and Camier* (Calder, 1974)
David Bellamy, *Bellamy's Ireland: The Wild Bogland* (Christopher Helm, 1987)
David Blackbourn, *The Conquest of Nature* (Pimlico, 2017)
Enid Blyton, *Five Go to Mystery Moor* (Hodder and Stoughton, 1954)
Boswell and Johnson, *Journey to the Western Isles of Scotland* (1785; Penguin Classics, 1984)
Emily Brontë, *Wuthering Heights* (Penguin Classics, 2009)
Robert Chambers, *The Book of Days: A Miscellany of Popular Antiquities* (Andesite Press, 2015)
Donal Clarke, *Brown Gold: A History of Bord na Móna and the Irish Peat Industry* (Gill and Macmillan, 2010)
Iain Crichton Smith, *On the Island* (Gollancz, 1979)
John Feehan, *The Bogs of Ireland* (University College Dublin, 1996)
Alan Gailey and Alexander Fenton, *The Spade in Northern and Atlantic Europe* (Ulster Folk Museum, 1970)
Robert Garioch, *Collected Poems* (Polygon, 2004)
P.V. Glob, *The Bog People* (Faber and Faber, 1965)
Lewis Grassic Gibbon, *Sunset Song* (Canongate, 2008)
Neil M. Gunn, *Young Art and Old Hector* (Souvenir Press, 1976)
Seamus Heaney, *Open Ground: Poems 1966–1996* (Faber and Faber, 1996)
James Hunter, *Set Adrift Upon the World* (Birlinn, 2015)
Roger Hutchinson, *The Soap Man: Lewis, Harris and Lord Leverhulme* (Birlinn, 2005)
James VI and I, *Daemonology* (Aziloth Books, 2012)
Suzanna Jansen, *Het Pauperparadijs* (Balans, 2011). No edition in English; see website: www.suzannajansen.nl/index.php
Patrick Kavanagh, *Collected Poems* (Penguin, 2005)
Mart Laar, *The Forgotten War* (Grenader, 2007)
Halldor Laxness, *The Fish Can Sing* (Panther, 2001)
Dane Love, *Scottish Covenanter Stories* (Neil Wilson Publishing, 2012)
Hugh MacDiarmid, *The Hugh MacDiarmid Anthology* (Routledge and Kegan Paul, 1972)
Norman MacCaig, *Collected Poems* (Chatto and Windus, 1998)
Dr Donald Macdonald, *Tales and Traditions of the Lews* (Birlinn, 2000)
Hector Macdonald, *A View from North Lochs* (Birlinn, 2007)
Alexander Mackenzie, *The Prophecies of the Brahan Seer: Coinneach Odhar Fiosaiche* (Constable, 1984)
Louis MacNeice, *Selected Poems* (Faber and Faber, 1988)

John McGrath, *The Cheviot, the Stag and the Black, Black Oil* (Bloomsbury Publishing, 2015)
Thomas Manson, *Humours of a Peat Commissioner* (Shetland News, 1918)
Geert Mak, *An Island in Time: The Biography of a Village* (Vintage, 2010)
Ann Michaels, *Fugitive Pieces* (Vintage, 1998)
John Montague, *Selected Poems* (Bloodaxe, 1990)
Flann O'Brien, *The Dalkey Archive* (Flamingo Modern Classics, 1993)
Flann O'Brien, *The Poor Mouth* (Picador, 1975)
Sean O'Casey, *Juno and the Paycock* (Macmillan, 1927)
David Profumo, *Bringing the House Down* (John Murray, 2006)
James Rebanks, *The Shepherd's Life* (Allen Lane, 2015)
Maurice Riordan, *A View from the Loki* (Faber and Faber, 1995)
T. A. Robertson, aka 'Vagaland', *Laeves fae Vagaland* (Shetland Times, 1952)
Tim Robinson, *Connemara: Listening to the Wind* (Penguin, 2007)
Ian D. Rotherham, *Peat and Peat Cutting* (Shire Library, 2011)
Karin Sanders, *Bodies in the Bog and the Archaeological Imagination* (University of Chicago Press, 2009)
Simon Schama, *Embarrassment of Riches* (HarperCollins, 1987)
Shakespeare, *King Lear* (Oxford University Press, 2013)
R. S. Thomas, *Collected Poems 1945–1990* (Phoenix, 1993)
Michel Tournier, *The Erl-King* (Harper King, 1972)
Nicholas Wachsmann, *A History of the Nazi Concentration Camps* (Farrar, Strauss and Giroux, 2015)
Dr Ali Whiteford, *Lewis Chemical Works* (Stornoway Trust, date unknown)

Journals and other media

Luka Bloom, 'Bogman' (song)
Archives department, Na h-Eileanan Siar
Scottish Islands Explorer, November/December 2016 (Warner Group)
Stornoway Gazette, various issues
West Highland Free Press, various issues

A Spadar Gives Thanks ...

In the course of writing this book, I encountered the term 'spadar', which is the Irish for 'wet, heavy turf' that smoulders rather than burns on the fire. *Brewer's Dictionary of Phrase and Fable* went on to inform me that the word was also 'used in Hiberno English to denote a useless, cloddish person'. As such, it seems the perfect term to describe me as I come to express my thanks to all those involved in the creation of this book. Names escape me. Thoughts slip away. However, a large number of people still come to mind. Many of them appear in the pages before this, but there are others who are not mentioned and it is to them that I would mostly like to express my gratitude. They include those who gave me a warm welcome and hospitality when I visited the south of Scotland: Ted Cowan, David Bartholomew, Mike Brown, Hugh MacMillan, Hugh Bryden, Chrys Salt and Richard MacFarlane (the last-named pair even gave me the use of their home). I am grateful for their kindness and friendship, all of them. They are truly wonderful, kind people.

In Ireland, I was grateful for the assistance of two of my former pupils, Maureen Mackinnon and Alyth MacCormack. Maureen introduced me to her friends – including Breandan and Maire Ferriter – in An Cheathrú Rua, County Galway. Alyth's role was of equal importance. Without her, I would never have met Kevin Conneff, whose presence and good company meant so much to both me and this book. I have an extra reason to thank my friends and former colleagues, Angus and Mary MacCormack, for their role in encouraging and fostering the talents of their daughter. She has returned them fourfold. Thanks to you all. My gratitude, too, goes to Janice Fitzpatrick Simmons and her son James for their help along the way. I was also delighted to make the acquaintance of Dr Liam Campbell and everyone I met at the Song of the Bog Conference at the Lough Neagh Discovery Centre. Iain MacAulay is owed a great debt for taking me along with him to that occasion. It was a wonderful place to be.

In the Netherlands, I was delighted to once again receive the support of various individuals. They include the Bosch family, Roel, Marleen and Aleid. (This book is dedicated to the first two for good reason.) I would also especially like to thank Elleke Bal, Suzanna Jansen (whose book is surely overdue being published in this country), Karel Leenders, Kerst Huisman – author of *Opstand in de Turf* (*Revolt in the Peat*) and the staff at Openluchtmuseum It Damshûs, who act as the guardians for the legacy of Ferdinand Domela Nieuwenhuis and the turf cutters more generally. Once again, I also wish to thank Huib Stam, whose courtesy and generosity is inspiring to a fellow writer. I also encountered a few people with Danish and German connections, who were incredibly helpful. These include a former colleague, Graham Kerr, and a past pupil, Roddy Walker. There were also others with Finnish and Estonian links who provided support. Gustav, Kivi, Sanna, Paavo and Annu – you know who you are. Adalsteinn Ásberg Sigurdsson and Örlygur Kristfinnsson made valuable contributions from Iceland.

But most of all, there are my fellow Scots. They include the financial and other forms of support from Creative Scotland, Shetland Amenity Trust, Sabhal Mòr Ostaig and Aros Centre in Skye. (I don't know how – to turn round the words of Tennessee Williams – I have come to 'depend on the kindness of Sgiathanaich'.) I have also obtained more than my share of support from my fellow writers and islanders in Lewis, Harris, Shetland and elsewhere, with names like Iain Gordon Macdonald, Donald Anderson, Alasdair and Deirdre Roberts, Donald Michael Macinnes,

James Andrew Sinclair, John Urquhart, Dr Catriona Macleod (who made the very helpful suggestion of a book), John Neil Munro, Mairi MacAulay, the late and much mourned Norman Maclean, Dolina Maclennan (who introduced me to the folksinger, Ian McCalman) and my brother Allan all coming immediately to mind when I write these words, though there are many others, including those I mention in the book. Once again, I would like to thank Kathleen, Karen and the rest of the library staff in both Shetland and the Western Isles. Their assistance is always invaluable. The same is also true of Brian Smith and the rest of the archive staff in Shetland Museum. I am especially grateful for the help of Angus Johnson in bringing a particular book to my attention.

Whether they are aware of it or not, these people are a major part of a team – one that is dependent on another four individuals, my agent Judy Moir and my two editors, Mari Roberts – who put up with all my errors with extraordinarily good grace and humour – and the wonderful Julie Bailey at Bloomsbury. Finally, there is the woman, Maggie Priest, with whom I share my life. In the alchemy and magic they produce, they resemble those that extracted *dargveen* and *baggel* from the sea waters that washed around the Netherlands, transforming them into a fuel that provided heat and light and flame.

Tapadh leibh uile.
Donald S. Murray, March 2018

Index